T0075920

Ice Ages

What causes Ice Ages? How did we learn about them? What were their effects on the social history of humanity? Allan Mazur's book tells the appealing history of the scientific "discovery" of Ice Ages. How we learned that much of Earth was repeatedly covered by huge ice sheets, why that occurred, and how the waning of the last Ice Age paved the way for agrarian civilization and, ultimately, our present social structures. The book discusses implications for the current "controversies" over anthropogenic climate change, public understanding of science, and (lack of) "trust in experts." In parallel to the history and science of Ice Ages, sociologist Mazur highlights why this is especially relevant right now for humanity. *Ice Ages: Their Social and Natural History* is an engrossing combination of natural science and social history: glaciology and sociology writ large.

Allan Mazur is a professor emeritus in the Maxwell School of Citizenship and Public Affairs at Syracuse University. He has also worked in the aerospace industry. He is a Fellow of the American Association for the Advancement of Science. He is interested in biosociology and conflicts over science, technology, and the environment. He is author or co-author of over 200 academic publications and ten books, including *Biosociology of Dominance and Deference* (2005), *Global Social Problems* (2007), and *True Warnings and False Alarms: Evaluating Fears about the Health Risks of Technology, 1948–1971* (2010).

'Allan Mazur takes us on a fascinating journey through two million years of Earth history and human history, linking the two through a lucid description of the great Ice Age fluctuations in climate. This is a book for all readers interested in our shared human career, and in how the dynamic surface of the Earth has influenced that career through the ages.'

Peter Bellwood, Australian National University

'Allan Mazur gives us a masterful exemplar of the history of science. He shows specialists from several discipilines and nonspecialists with just a modicum of science how diverse paths of inquiry over recent human history have revealed the details of prehistory going far back into geological time. He shows us how more detail is known than might have been imagined when the scientific work began in the 18th century. Not since Simon Winchester's *Krakatoa* has the science of geology been so absorbing! More importantly, Mazur shows both how ice ages – large and small, long and short – and their endings have changed human history, and how our short-sightedness about their causes and effects is going to change future human history, for the worse . . . unless the right people learn the lessons of this book.'

Alex Rosenberg, Duke University

'Living on a warming planet, we struggle to imagine that it was periodically covered by vast sheets of ice. Allan Mazur, a master of calm, companionable, and often humorous prose, guides us through the various efforts humans – plucky survivors of the Pleistocene – have made to understand the Earth as well as their transformative and, it now turns out, damaging presence on it. An impressive synthetic effort, blending science and cultural history, Mazur's excellent *Ice Ages* gives us the tools necessary to participate knowledgeably in debates about climate disruption.'

Christoph Irmscher, Indiana University, author of *Louis Agassiz: Creator of American Science*

'This absolutely fascinating book weaves together the complicated strands of human endeavor that led to the great scientific discovery of ice ages on Earth. It should be read by everyone interested in the current pressing problem of global climate change, both natural and human induced.'

George Denton, University of Maine

Ice Ages

Their Social and Natural History

Allan Mazur
Syracuse University

CAMBRIDGE
UNIVERSITY PRESS

University Printing House, Cambridge CB2 8BS, United Kingdom

One Liberty Plaza, 20th Floor, New York, NY 10006, USA

477 Williamstown Road, Port Melbourne, VIC 3207, Australia

314–321, 3rd Floor, Plot 3, Splendor Forum, Jasola District Centre, New Delhi – 110025, India

103 Penang Road, #05–06/07, Visioncrest Commercial, Singapore 238467

Cambridge University Press is part of the University of Cambridge.

It furthers the University's mission by disseminating knowledge in the pursuit of education, learning, and research at the highest international levels of excellence.

www.cambridge.org
Information on this title: www.cambridge.org/9781316519400
DOI: 10.1017/9781009023566

First published 2022

Printed in the United Kingdom by TJ Books Limited, Padstow Cornwall

A catalogue record for this publication is available from the British Library.

Library of Congress Cataloging-in-Publication Data
NAMES: Mazur, Allan, author.
TITLE: Ice Ages : their social and natural history / Allan Mazur, Syracuse University.
DESCRIPTION: Cambridge, UK ; New York, NY : Cambridge University Press, 2022, | Includes bibliographical references and index.
IDENTIFIERS: LCCN 2021025793 (print) | LCCN 2021025794 (ebook) | ISBN 9781316519400 (hardback) | ISBN 9781009010443 (paperback) | ISBN 9781009023566 (epub)
SUBJECTS: LCSH: Paleontology–Pleistocene. | Paleoclimatology–Pleistocene.
CLASSIFICATION: LCC QE741.2 .M39 2021 (print) | LCC QE741.2 (ebook) | DDC 560/.1792–dc23
LC record available at https://lccn.loc.gov/2021025793
LC ebook record available at https://lccn.loc.gov/2021025794

ISBN 978-1-316-51940-0 Hardback

To Polly, who gave me everything.

CONTENTS

PREFACE

This book contains little original information. Its contributions lie in my selection and assembly of work by prior scholars into an overview of how, over little more than two centuries, we came to understand the history of life and of the Pleistocene Ice Age. Humans (the genus *Homo*) first came on the scene during the Ice Age, and the sole remaining species of that genus was *Homo sapiens* by its end. Recession of the great northern ice sheets was a precondition if not an essential stimulus for nearly all surviving human communities to switch from hunting and gathering to an agrarian lifestyle, some later elaborated into the earliest civilizations and only recently into industrial society. This is human history on the grandest scale. It is remarkable that many Americans do not believe essential features of the story.

Combining so many disparate strands and sources to weave this narrative, one confronts among other problems that of different measurement units. There are British derived yards, feet, and miles, still used in the United States, which is resistant to adopting the more logical metric system. I have sometimes included both systems, one in parentheses, but as often simply used one or the other, depending on context, when I think it intelligible to readers.

I appreciate the help and advice of Calvin Barrett, Peter Bellwood, Richard Buttny, George Denton, Arwyn Edwards, Meredith Berman Ellis, Christoph Irmscher, Linda Ivany, Ross MacPhee, Peter Richerson, Alex Rosenberg, Martin Rudwick, Liz Williams, my daughters, Rachel Mazur and Julie Mazur Tribe, and my brother, Melvin Mazur. Any errors remaining are my own. Thanks to the people at Cambridge University Press: Sapphire Duveau, Michale Goldstein, Sarah Lambert, Matt Lloyd, Sindhuja Sethuraman, and Diana Witt. John and Matt, thank you for past and future talks about rocks and fossils. Max and Wren and Marlo, thank you most of all.

1 IN THE BEGINNING

> Twenty-thousand years ago, the earth was held in thrall by relentlessly probing fingers of ice – ice that drew its power from frigid strongholds in the north, and flowed southward to bury forests, fields, and mountains. Landscapes that were violated by the slowly moving glaciers would carry the scars of this advance far into the future. Temperatures plummeted, and land surfaces in many parts of the world were depressed by the unrelenting weight of the thrusting ice. At the same time, so much water was drawn from the ocean to form these gargantuan glaciers that sea levels around the world fell by 350 feet, and large areas of the continental shelf became dry land.
>
> *Imbrie and Imbrie (1979: 11)*

The Pleistocene epoch, or Ice Age, is today dated from 2.6 million to 12,000 years ago, a period marked by repeated advances and retreats of great ice sheets. When glaciation was last at its maximum, little more than 20,000 years ago, the area that is now New York was covered by one or two miles of ice and snow. If there had been skyscrapers in Manhattan, most would have been buried, with only a few peeking out like islands in a white sea.

The Pleistocene Ice Age and the evolution of the genus *Homo* occurred simultaneously. Usually they are described separately, but as a sociologist and historian of natural history, I am treating them together. I will tell how savants/scientists of the past two centuries came to our present understanding of the ice ages, their causes and consequences – in that sense giving a history of scientific understanding. I also describe the changes themselves, how ice repeatedly came and went and how evolving humans responded to these conditions. Those of the Neanderthal type adjusted well to cold climates, having a very long tenure in Eurasia until replaced by modern people.

Of several *Homo* species whose fossils are known, there remains a single species that by the epoch's end occupied the presently inhabited world. Roughly 10,000 years ago, as the ice receded, surviving human communities switched from hunting and gathering to an agrarian lifestyle. Within a few thousand years more, some of these communities developed into the world's earliest civilizations, independently in the Middle East, China, and India, and somewhat later in the Americas. This is human history on the grandest scale. The story is credible only because we finally discarded traditional beliefs that were unquestioned for a thousand years.

In the eighteenth century, European and American savants had no notion that life had a far longer history than told in the Bible or that humans were late arrivals on the scene. They believed that Earth's surface was shaped primarily by a Great Flood and occasional volcanos. Sculpting by vast ice sheets was unimaginable. *Ice Ages* is the story of how, over little more than two centuries, an exciting series of discoveries, errors, controversies, and reinterpretations led to the accumulation of solid evidence for what is now taught in every university, a fascinating story of scientific discovery about the most recent Ice Age and the history of human life.

The last Ice Age lasted for a time barely imaginable before two centuries ago. Archbishop James Ussher (1580–1656) had famously calculated that Creation occurred in 4004 BC. It follows that Earth was 5,863 years old in 1859, the year Darwin published *On the Origin of Species*. This biblical dating, often ridiculed today, was not unusual or out of line with other estimates based on similar assumptions. The Judaic calendar, still in use for religious purposes, equates the secular year 1859 with the Jewish year 5619, counting from the beginning.

The Bible's enumeration of the generations since Adam – the who begat whom of the King James Version – gives sufficient information to tally the elapsed time from Adam's first breath to the birth of Noah as 1,056 years. Noah was 600 years old when water flooded the planet (Genesis 7:6). That dates the Great Deluge at 1,656 years after Creation, or 2348 BC, according to Archbishop Ussher's count.

The longevity of the progenitors is striking. Adam lived to the age of 930, Methuselah to 969, and Noah to 950. Only 126 years separate the death of Adam from the birth of Noah. Nearly all the several progenitors were alive at the same time. They could have come together for the weekly day of rest, exchanging stories about times past.

I confess to astonishment that so many in the United States today think that the Bible with its 6,000-year-old universe is literally true. Yet surveys show a third of Americans believing that it is, word for word (Mazur 2008).

Even in centuries past, many scholars did not read biblical texts with that exactitude. Apart from the fact that many people could not read at all, there was no original Bible to read, nor even trustworthy copies of an original, nor any surviving document older than a few centuries before the Christian era. The fourth-century Catholic Church was sufficiently concerned about inconsistencies and errors among the various biblical manuscripts in circulation that Pope Damasus I commissioned his secretary, Jerome, to produce a standardized Latin translation of the New Testament. Working for twenty years, Jerome compared numerous texts in Hebrew, Aramaic, Latin, and Greek before settling on his own version, later known as the Vulgate Bible, the first book printed by Gutenberg and, to this day, the official text of the Roman Catholic Church. Some passages that made the cut might have strained Jerome's own monotheistic credibility, such as the account in Genesis (6: 1–4) of when the Nephilim were on Earth, when the sons of God went into the daughters of humans, who bore children to them: "These were the heroes of old, warriors of renown."

After Jerome produced the Vulgate Bible from discrepant sources, there remained inconsistencies. For example, Genesis opens with two successive accounts of Creation, one occurring in seven days, the other about the Garden of Eden. In the first story, humans appear on the sixth day, after plants and animals. In the second story, God creates a man (Adam) "from the dust of the ground," then creates plants in Eden where the man can dwell, then forms every animal and bird, and finally creates a woman.

Some traditionalists see the Bible's two stories of creation as a telescopic narrative, with the opening account giving the "big picture" while the story of Eden narrows the focus. Adam and Eve's tale is so engagingly different from the impersonal catalog of seven days that casual readers may not notice their contradictions. In the seven-days story, all vegetation including seed plants and fruit trees is made on the third day. All sea creatures and flying birds are made on the fifth day. All land animals from cattle to creeping things are made on the sixth day, and afterward God makes humans – male and female – to rule over these fish, birds, and animals, and to use the plants for food.

In the second story, Adam comes first, "when no plant of the field was yet on the earth." Then plants are created in Eden. Then every animal and bird is formed. Finally, Eve is made from Adam's rib.

Was the first man created before plants and animals and birds or afterward? Did birds appear before land animals or at the same time? Early readers within both Hellenistic and rabbinic traditions recognized these inconsistences. The sequences agree on only two points: (1) vegetation preceded animals and birds and (2) the first woman was created at the end of the process. There is little correspondence between either of the biblical sequences and our modern understanding of life's history, but ancients did not need that knowledge to recognize that the two accounts are not logically consistent.

After Cain slew Abel, he went to the Land of Nod to the east of Eden and there married. Where did his wife come from? Was she his sister? Noah brought into his ark a pair of each kind of animal (6:17–22), but elsewhere we are told he brought in seven pairs of all clean animals and birds (7:1–5). The Flood lasted forty days and nights (7:4, 12, 17), but elsewhere "the waters swelled on the earth for 150 days" (7:24). Other passages tell of the ark coming to rest on the mountains of Ararat five months after the onset, with Noah and his family and the animals emerging after a year of confinement (7:11, 8:4, 14–16).

The Catholic position has always been that the Bible must be understood in accord with Church interpretation. (This sometimes changes, as when the Vatican accepted the sun-centered solar system and now accepts the evolution of life.) Not until the Protestant Reformation did breakaway denominations emphasize that whatever the Scripture said was true and that each literate person could read it for himself, without relying on interpretations pronounced by the corrupt Vatican hierarchy. Nonetheless, many Protestant scholars, while accepting religious and moral precepts of the Bible, had long disregarded its natural history as incorrect or implausible in detail. After all, the first five books of the Bible, which Jews call the Torah and Christians the

Pentateuch, were traditionally said to be authored by Moses, yet they include events before his birth and after his death. How could that be? (Perhaps by divine revelation.) As Old Testament scholar Richard Friedman puts it, "[T]he primeval history is barely capable of being considered from the point of view of historicity, given its conception of a finite universe surrounded by water, a talking snake, 'sons of God(s)' having relations with human women, a box [ark] containing the whole of animal life, and simultaneous creation of languages" (2003: 620).

There is little reason to think that many eighteenth-century savants – the word "scientists" was not yet in use – believed Archbishop Ussher had truly dated the beginning of the universe. Some even wondered whether Earth or the universe had a beginning at all. Apart from the Genesis account, there was no reason to assume a creation event or a worldwide deluge. Perhaps Earth always existed pretty much as it was, without dramatic interludes, and always would.

The Scottish Enlightenment philosopher-farmer-physician James Hutton (1726–1797), a deist often called the "father" of modern geology, ignored biblical stories about a Creation and Flood, instead postulating that the past history of the globe must be explained by processes that we can see happening now, or to have happened recently. Hutton did not only speculate at his desk but, like others at the time, went into the field, examining the landscape first-hand, which became the essential investigative model for geologists of the next two centuries.

Hutton saw no compelling reason to give Earth any birth date and believed sufficient time had elapsed for commonplace forces such as erosion and deposition, punctuated by volcanic eruptions and earthquakes, to create the ever-changing landscape. He explained that marine fossils were sometimes found far from the sea because a gradual uplift of land had caused recession of the shoreline, so that animals once living in the ocean were now buried high and dry. For Hutton, the world was in a steady state, changing gradually with "no vestige of a beginning, no prospect of an end" and therefore no significant history of great events or cataclysms. It was a world, he thought, eternally habitable by humans.

On the other hand, for savants who did believe in a finite Creation, whether or not as told in Genesis, there was no reason to think it occurred a vastly longer time before there were humans on the scene. Apart from Bible stories, how would anyone know?

It is often said that two great revolutions transformed our sense of place in nature. First was the Copernican Revolution, which demoted Earth from the center of the universe to one among other planets revolving around the sun; this solar system was eventually recognized as one locale in an enormous galaxy, which itself was one among billions of galaxies. The second revolution was Darwinian, placing *Homo sapiens* within the tree of life, still special in our own way but at base simply one of endless forms, most beautiful and wonderful, that have been, and are being, evolved.

The preeminent historian of paleontology, Martin Rudwick, points to another revolution between the Copernican and Darwinian revolutions, equally important for our place in nature but less recognized: enlarging the timescale of Earth and, by implication, of the universe.

> In earlier times, most people in the West had taken it for granted that the world had started, if not precisely in 4004 BC, then at some such point in time, only a few millennia ago. After this revolution it became equally commonplace to accept that the Earth's timescale runs at least into millions of years, if not billions ... The "young Earth" of the traditional picture was also an almost wholly human Earth. Apart from a brief opening scene ... it was a human drama from start to finish, from Adam through to some future Apocalypse at the end of the world. In contrast, the "ancient Earth" first discovered and reconstructed by early geologists was largely non-human because it was almost completely pre-human: our species seemed to have made a very late appearance on the world stage. (2014: 2)

Realizing the long time span gave Darwin the time needed for natural selection to work. More importantly for present purposes, it was the foundation for a history of Earth itself – changing land masses, uplift and erosion of mountains, long-term changes in climate, and the formation of continent-wide ice sheets.

We begin with an account of how European savants came to realize that Earth is far older than 6,000 years, and that humans were late arrivals on the scene. I hope the reader will enjoy this story as much as I did in bringing it into narrative form.

2 "BURSTING THE LIMITS OF TIME"*

Near the rim of Grand Canyon are national park bookstores, some selling a beautifully illustrated volume explaining that the great chasm was created less than 6,000 years ago by Noah's Flood. While paging through it and finally deciding against paying the considerable price, I fell into conversation with the salesclerk who volunteered that he believed it was true. I did not argue, knowing it would be fruitless, but chalked it up as one of the few instances I've ever met a Christian fundamentalist. There are a lot of them in America but not in university circles, nor, for that matter, in the neighborhoods of Chicago where and when I grew up.

Today's scientific view is that Earth and life had discreet beginnings and subsequent histories of vast duration, revealed not by the Bible but by physical evidence. It is this enlargement of the time scale that Martin Rudwick ranks in importance with the Copernican and Darwinian revolutions. This monumental change occurred quickly, within roughly the half century after the French Revolution. How did it happen?

There were at the time good practical reasons to better understand the Earth's crust. The Industrial Revolution, which began in eighteenth-century Britain and spread quickly to Europe and America, increased demand for coal and ores. Several European governments founded mining academies, their graduates mapping mineral deposits

* This title is taken from Rudwick (2008), who took it from Georges Cuvier, *Researches on Fossil Bones*, 1812.

below the landscape to guide the opening of new quarries and mine shafts. These mining engineers interpreted rock layers structurally rather than chronologically. Still, this was the beginning of an intellectual expansion of nature's time scale. Rocks and fossils were the primary evidence. But evidence is not interpreted in a vacuum. Important context was provided by two cultural movements of the times, the Enlightenment and the Romantic.

The Age of Enlightenment

Isaac Newton's *Principia* (1687) inspired the Enlightenment and the new spirit of science: Do not trust traditional authority; seek truth through logic, observation, and experiment. Do not believe what churches or ancient writers taught, or what priests and kings say, but put knowledge to the test.

If Newton could, by logical deduction, explain orbital and terrestrial mechanics as an integrated whole and predict motions verified by observation, then why couldn't logic and systematic observation explain other aspects of the natural world, not only physics and chemistry, but geology and biology as well? Human history might be reconstructed using evidence from the past, as in excavations recently begun at Pompeii and Herculaneum, the Roman cities buried in the catastrophic eruption of Mount Vesuvius in AD 79. Could natural history be reconstructed the same way?

Even the Bible was examined through a rationalist lens. Literary scholars of the eighteenth and nineteenth centuries developed methods of textual analysis focused on such questions as whether a single author did indeed write all the works attributed to Shakespeare. Their method, very briefly, is to compare themes and writing styles of the different works on the assumption that authors may be recognized by their unique and consistent forms of expression, grammar, choices of words, and punctuation. In Germany, scholars applied the same method to the Bible not to undermine belief but to gain a better understanding of this holy text.

Compare the two versions of Creation: Not only do they contradict one another in sequencing the appearance of life forms. They also differ in overall style, one a log of seven days, the other a tale that a bard might tell about specific people, Adam and Eve. They differ in referring to the deity. The seven-days version speaks impersonally of "God" (in Hebrew *Elohim*). In the Adam and Eve tale, God is called by his personal name, Yahweh. In the seven-days account, the words used for creation are derivatives of one Hebrew root; in the Adam and Eve account they are derived from a different root. There is a compelling case that the two passages were written by different authors.

After three centuries of research, many biblical scholars (though not fundamentalists) agree that the Torah is a composite of previously separate source documents that describe the same events, such as the Creation and Flood, though with different details. Adam and Eve's story, which speaks of Yahweh, is the opening portion of the "Yahwist," or J, document (for *Jahwist*, as German scholars spell it). The seven-days story opens the "Priestly," or P, document, because of its exceptional interest in priestly issues.

On stylistic evidence, the entire text can be parsed into separate J and P documents, and a few other sources (see Friedman 2003). Each source is consistent within itself, though differs from others in detail. The J and P documents are hypothetical constructs. No fragment exists, and their authors are unknown. The late Professor Harold Bloom of Yale provocatively argued that J's author was a high-born lady in King David's court.

The source documents were compiled into the full *Torah* by an editor or editors (redactors). Judging from textual clues, most of this compilation occurred during the few centuries before the Christian era. Interweaving different stories about the Creation, the Flood, etc., each with different details, produced inconsistencies in the full text. For rationalist scholars, such analyses cast doubt on the Genesis account of historical events, fueling a religious movement toward deism.

Enlightenment geologists, whether deists or orthodox Christians, tried to understand Earth in a rational framework. What had earlier been casual observations of salient formations like the White Cliffs of Dover, or the "layer cake" stacking of distinct mineral strata (as at the Grand Canyon), were now catalogued carefully, where they could be seen in cliff faces, river cuts, quarries, and mine shafts. This too suggested interpretations at odds with Genesis. The Dover Cliffs, rising 350 feet (ca. 100 m) above the sea, were seen from microscopic examination to be an aggregate of tiny marine skeletons, apparently accumulated over a long period of time. Deist geologists like Hutton, freed of Scriptural constraint, believed such large accumulations of sedimentary material, and the neat stacking of strata, must have occurred gradually, over a very long time, far longer than 6,000 years.

By the late eighteenth century, savants took it as much for granted that the Earth was almost *inconceivably* ancient as their predecessors a century earlier had assumed that its history could be measured in mere millennia. The many savants who regarded themselves as Christian believers found this extended timescale no more problematic or disturbing than their unreligious contemporaries. Biblical scholars had recognized the ambiguity of the key word "day" in the Genesis narrative, long before any evidence derived from the natural world began to throw doubt on the traditional and common-sense assumption that it denoted an ordinary day of twenty-four hours (Rudwick 2014: 100).

Even as the belief in six days of Creation fell away, Genesis retained some hold, segueing from *the* Creation to successive creations as new kinds of life appeared in higher strata, and from Noah's Flood to successive floods, each a catastrophe if on a smaller scale, to explain discontinuities in the fossil record.

The Romantic Era

The Enlightenment, perhaps overly rational, begat the Romantic Era in reaction, influencing art, literature, music, and intellectual activity. Romantics stressed the importance and beauty of feelings and emotions, the sublimity of God's creations, untrammeled by human artifice, but also the aesthetic of Classical and Gothic constructions, now in ruins, entangled in vegetation. This was particularly important for views of the landscape, outside of the ugly but growing cities.

In England, eventually across Europe, landscape gardening moved toward an idealized view of nature, replacing the formal, rationalist, symmetrical French style, epitomized by the grand gardens of Versailles. At rich manor houses, the English style sought a pastoral feel, including open lawns, groves of trees, a lake or pond, and, perhaps as a "conceit," a small Greek temple or newly constructed ruin. For the merely well off, the English garden featured pathways through a thickly planted mix of flowers and shrubs, in contrast to the uniformly spaced and geometrically artificial flower-plots of the French garden.

Places that had previously been seen as dangerous wilderness, the haunts of wild animals, savages, thieves and perhaps witches, now became scenic attractions. Hiking and sight-seeing in the Alps became a

Figure 2.1 Left: *Wanderer above the Sea of Fog* by Caspar David Friedrich (1818); Right: *Louis Agassiz at the Unteraar Glacier* by Alfred Berthoud (1881) in the Library of the University of Neuchâtel

vacation activity, in which savants fully participated. Caspar David Friedrich's 1818 painting, *Wanderer above the Sea of Fog*, epitomizes to me the European gentleman surveying nature from atop an Alpine peak (Figure 2.1). The wanderer could have been the Swiss trekker Louis Agassiz a few years later, on one of his many ventures into the Alps and Jura mountains, visualizing how the landscape was, in an earlier epoch, covered by a vast sheet of ice.

This new appreciation of natural landscape contributed to rising interest in natural history, not only among savants but the public too. Nature became recreational as people at leisure walked or rode to scenic viewpoints, collecting natural objects along the way, perhaps minerals or other interesting stones, flowers and leaves. Painting wildflowers was in vogue (Kramer 2002). Outings were a good opportunity for bird watching. In Upstate New York, still a wilderness frontier, James Fenimore Cooper (1789–1851) wrote his popular "Leatherstocking" novels, including *The Last of the Mohicans*, expressing the noble virtue of American Indians. Traveling across the new nation, John James Audubon (1785–1851) shot and then painted his collection of North American birds.

An earlier tradition of "cabinets of curiosities" revived. These were collections of almost any objects that seemed interesting, whether natural or manufactured curios, which could be displayed for visitors, a mark of the owner's intellectual stature. Fossils were prominent among these displays, increasingly so with the vogue in natural history. Clergymen, their workloads relatively light during the week, became avid hikers and collectors of natural objects, which could be picked up from the ground at no cost. These were arranged for display on no general principle beyond the pure pleasure of seeing the mélange of objects.

Practical Engineering

By the nineteenth century, Europe and the United States were deeply into the Industrial Revolution, with factories in the Midlands and New England replacing cottage industry to manufacture yarn and cloth from cotton. Soon came the development of railroads and iron and steel, fed by coal and ores. There was strong economic incentive to locate and mine these minerals from quarries or deep shafts.

Casual observers of cliff sides had always noticed their usually horizontal layers, or "strata," marked by different colors and comprising different kinds of rock. Now there was good reason to study them more closely, to identify likely sources of valuable resources. Sometimes the rock within a stratum was highly jumbled. At other times, the strata seemed gradually laid down with sharp boundaries between adjoining layers; these were called "conformities," indicating an orderly superposition of one layer atop another.

Higher layers were often sedimentary formations like sandstone, limestone, or shale. Such deposits were known to occur in bodies of water, as floating particulates, or calcite remnants of shells and fish skeletons, settled to the bottom, sometimes forming enormous accumulations. In Hutton's terms, this was the build-up of mineral deposits by gradual deposition. If the body of water later receded or dried up, the surface would be exposed to the weather and smoothed by erosion. If water returned to the area, there would be a new layer of sedimentary deposition, its weight compressing the strata below. These orderly conformities did not look anything like one would expect if rocks had been deposited in a short period of time by turbulent waters of a great flood. There were also nonconforming layers, in which rocks were jumbled, but these might underlie a smooth stratum, again arguing against one sudden, turbulent deposit.

There were also non-sedimentary layers, apparently hardened lava from once active volcanos, and sometimes these intruded into fissures in a sedimentary layer, indicating that the volcanic eruption occurred after the sediment had been laid down. Thus developed a sense of chronology in the build-up of layers, with occasionally chaotic interruptions.

Engineers mapping these formations were not especially interested in philosophical implications, but in the practical goal of locating coal seams and valuable mineral deposits. In the process, almost incidentally, they found in different places that the deepest rocks, presumably the oldest, had no fossils. At least as early as 1756, the German mineralogist Johann Lehmann (1719–1767), studying mining in Prussia, proposed a threefold division of rocks into (1) *Primitive*, including granites and gneiss in which there was no fossils, therefore antedating the creation of life; (2) *Secondary*, comprising fossiliferous strata of sandstone, shale, and limestone; and (3) higher *Alluvial* deposits of rock and gravel, due to local floods and the Great Deluge of Noah. (Subsequently, these divisions were altered, as when fossils were found in layers that Lehmann had regarded as Primary, and eventually replaced by modern names.)

Taken together, the great thickness of Primary formations hinted that Earth was in existence long before the appearance of life in the higher Secondary rocks. Furthermore, among the higher fossil-bearing rocks, no human bones were found much below the surface. After life first appeared, it seemed to have taken a long time for humans to come on the scene (Rudwick 2014).

Lehmann and other mining engineers interpreted these rock layers structurally rather than chronologically. Still, this was the beginning of an intellectual expansion of nature's timescale. Rocks and fossils were the primary evidence.

Fossils Were Alive

Prior to the eighteenth century, the term *fossil* was broader than its usage today, including crystals or minerals and interesting stones found on or dug from the ground. Some stones resembled animals, or parts of animals, though it might be hard to be sure if their features were often blurred from weathering. No matter, how could they be animal remains if they were made of stone? Some fossils looking like clamshells were found far from water, even on mountains, making it less plausible that they had once been alive at that place.

Two savants, Nicolaus Steno in Denmark (1638–1686) and Robert Hooke in England (1635–1703), are credited with a strong argument that these stones are remnants of living organisms. The crux was to present organic remains in an intermediate state of fossilization, usually plants or animals buried in lignite or brown coal, where they had partly changed to stone but not yet completely. Thus, the transition from life to stone was demonstrated. Once this was established, the process of fossilization was easily deduced: A dead creature quickly buried in mud would elude scavengers. As it decayed, it became porous, so grains from the mud would gradually fill vacant spaces and harden. With sufficient time, the creature would be transformed into a stone cast. If unearthed quickly, without prolonged exposure to the weather, intact molds may be in perfect condition, showing a "positive" cast of the creature, as well as its "negative" imprint in the surrounding matrix, the two fitting together like hand in glove.

Intriguingly, there were also fossils that did not look like anything known in the living world. Some speculated that these kinds had gone extinct, a strange idea to people raised on the notion that God created all species unchanging and enduring. Even discarding that theological baggage, it had to be granted that strange looking fossils might still reside in the depths of the ocean or in parts of the world where flora and fauna were largely unknown to Europeans or Americans. Perhaps ammonites and belemnites, abundantly found in the lower (hence older) Secondary layers but not at all in the higher formations, might simply mean that living specimens were unseen in the deep sea.

Crinoids, for example, are lovely fossils, resembling a flower on a stem (though actually an animal), abundant in lower Secondary strata and treasured by collectors (Figure 2.2). I often see stem parts in the Devonian shales of Upstate New York. Unlike anything then known to be alive, some savants thought them extinct until a living sample was dredged up in the West Indies. Today called "sea lilies," these animals live in tropical seas, though not nearly as abundantly. A more recent example is the coelacanth (*Latimeria chalumnae*), a fossil fish long thought extinct until in 1938 a living specimen was caught off the east coast of South Africa.

Figure 2.2 Left: *Agaricocrinus americanus*, a fossil crinoid from the Carboniferous. Right: Living crinoid.
Sources: Left, GNU Free Documentation license. Right, NOAA Photo Library – Flickr: expl5409, CC BY 2.0, Wiki Creative Commons

The fact of extinction was gradually established as more species were catalogued in the Americas and Europe's colonial possessions around the world, and as fossils that were discovered became larger, sometimes huge, making it implausible that if still alive they would escape notice.

Also, the blossoming discipline of comparative anatomy critically contributed evidence for extinction. Georges Cuvier (1769–1832), working at the *Jardin des Plantes* in Paris, was the foremost anatomist of his day, comparing living vertebrates with those dug up, which resembled but seemed not exactly like animals still in the wild. He identified a large skeleton dug in Paraguay as a giant prehistoric ground sloth, calling it *Megatherium*. Also, Cuvier confirmed that living elephants in Africa and India were different species, both differing from the fossil mammoths whose parts were found in northern Europe and America.

In 1799, near the mouth of the Lena River in Siberia, a nomadic hunter discovered a huge animal frozen in a block of ice, its features unclear. Over the next few years, as the body melted out, it could be seen as elephant-like. The large tusks were taken for sale, and most of the flesh was removed, either by hunters to feed their dogs or scavenged by wild animals. The account of Russian botanist Mikhail Adams (1808) tells of hearing about the find, reaching the remote site in 1806, and finding a nearly complete skeleton, later identified as a mammoth. The flesh was highly mutilated, but some remained attached. The head was covered by dry skin, the eyes preserved, the dried brain still in the cranium. Skin on the side where the animal had lain was thick and heavy, covered by red hair and black bristles.

Mammoths, if still alive, could barely remain unnoticed. Indeed, they *were* noticed by artists whose cave drawings in France, dated about 15,000 years ago, were rediscovered in the nineteenth and twentieth centuries (Figure 2.3). Their thick fur coats suggested that they lived in cold climates like in Siberia, not in the tropics like living elephants. Cuvier recognized and named another distinct fossil elephant, the "mastodon" of America, again with a furry coat suitable for a cold climate. As he identified more fossilized land mammals, large enough that it was implausible that they still lived unseen, the notion of extinction became widely accepted among savants, if not the cause of such extinctions. In his *Essay on the Theory of the Earth* (1813), Cuvier proposed that these now-gone species had been wiped out by periodic catastrophic flooding events. In this way, Cuvier became the most influential proponent of catastrophism, whether or not by biblical Flood, as counterpoint to Hutton's uniformitarian view.

Most of Cuvier's fossil mammals came from Alluvial deposits near the surface. These were sufficiently recent that the catastrophic

Figure 2.3 Mammoth (and ibex) drawn in the Rouffignac cave, Dordogne, France, ca. 15,000 years ago.
Source: Wikipedia

flooding causing their extinction might have been Noah's flood, though that was not Cuvier's own belief. These extinct animals differed from living animals, but not all that much.

Cuvier identified quadruped fossils from deeper deposits as reptiles, not mammals, and went so far as to suggest (incorrectly) that these layers were deposited when mammals had not yet appeared. "This was a first hint of a possible *history* of quadruped life; even perhaps a history that could be called 'progressive,' since human beings, universally regarded as the 'highest' of all mammals, seemed to have appeared even more recently, with no authentic fossils at all" (Rudwick 2014: 137). Some strange fossil reptiles appeared. Cuvier identified one as a *flying* reptile, naming it pterodactyl, or "wing fingered."

Fossil hunting was not only a pastime but a profession, aimed at satisfying the growing demand of collectors for strange remnants of life. Mary Anning (1799–1847), one of the most famous collectors, lived in the seaside town of Lyme Regis in Dorset on the English Channel. Coached in fossils by her father, who made cabinets in which affluent locals would display their specimens, she earned her living finding and selling fossils, some of them spectacular. At the age of twelve, she found in a cliff by the sea the first known ichthyosaur, a sea creature several meters in length. Among Anning's many finds were two complete plesiosaur skeletons, also large, and a pterodactyl (Pierce 2006).

Richard Owen (1804–1892) was England's later equivalent of France's Cuvier. Both were superb comparative anatomists, masters at reconstructing extinct animals from fossil remains. Owen is best remembered for delving further back in the history of land vertebrates, especially for coining the term *dinosaur*, meaning "terrible reptile (or lizard)." Owen based this new grouping on three genera whose parts were dug up in England: the carnivorous *Megalosaurus*, which Owen estimated at about nine meters long, the herbivorous *Iguanodon*, also several meters long, and the armored *Hylaeosaurus*, about five meters long. Statues based on Owen's notions of what these creatures looked like were constructed for the Great Exhibition of 1851 at London's new Crystal Palace (Figure 2.4). There could be little doubt that these animals no longer existed and did not resemble any animal now alive.

At different times during the Mesozoic, various kinds of dinosaurs appeared and then went extinct, replaced by other kinds of dinosaurs. The famous *T. rex* appeared late in the era, thriving in the western United States until that day 66 million years ago when a huge

Figure 2.4 Top: 1854 reconstruction in Crystal Palace Park, guided by Richard Owen, presents *Megalosaurus* as a quadruped, permission "Jennifer Crees/FCPD." Bottom: Modern reconstruction makes it bipedal, like all theropods, from Wiki Commons by LadyofHats

meteor or asteroid hit just offshore the Yucatan Peninsula, near today's Mexican resort area of Cancun. That ended nearly all dinosaurs excepting some smaller ones that had already evolved into birds. Perhaps their small size, relatively short egg-hatching period, warm blood, and ability to fly helped some of them survive that extinction, later diverging into today's 10,000 bird species (Brusatte, 2018).

Fossil Clocks

By the beginning of the nineteenth century, there had been decades of practical experience cataloging the rock strata underlying Britain and Europe to locate valuable deposits of coal and ores. In regions of England that had been surveyed, a thick white Chalk formation, produced by a lengthy sedimentation of small marine creatures, was the top stratum of Lehmann's Secondary rocks, lying just below Alluvial sands and gravels at or near the surface, its most famous outcrop the White Cliffs of Dover. The Chalk lies under both London and Paris, where it was quarried as a source of gypsum.

Early in the nineteenth century, Alexandre Brongniart (1770–1847), director of a porcelain factory outside Paris, found in his region belts of sandstone, clay, and limestone between the Chalk and the Alluvium. In much of England, these layers had eroded away, down to the Chalk, but not near Paris. (Layers above the Secondary's Chalk were later called "*Tertiary*," a name still in use, though "Primary" and "Secondary" are not.)

Cuvier, partnering with Brongniart, recognized that there were distinctive fossils in each stratum. Thus, he could distinguish one layer from another by comparing their fossils, which were as good as rock composition for identification.

In 1808, Brongniart and Cuvier displayed a map of the Paris region, colored to show the distribution of strata, which, being tilted, had exposed outcroppings at different places on the surface (see https://en.wikipedia.org/wiki/Alexandre_Brongniart). It appeared as if one could walk from the Chalk, one way or the other, across successively younger or older layers, assuming the concealing dirt and vegetation were removed. Furthermore, in some strata the fossils looked like shellfish known to live only in freshwater, while other strata had shells resembling those living in the sea. From this sequence, Brongniart and Cuvier inferred that the region has experienced repeated alterations of freshwater and marine conditions. Rudwick (2014: 132) cites their detailed analysis of the Paris Basin as the most influential example of how the static three-dimensional pictures of the engineers could be interpreted as a dynamic *history* of Earth's crust.

There were earlier geographical maps, one of the Eastern United States by Scottish-born William Maclure (1763–1840), who immigrated to America and made a fortune in business. In middle age, working alone, he crossed nearly every state, creating a map colored to show Primary, Secondary, and Alluvial exposures, presenting it in 1809, extended and revised in 1817 (see https://en.wikipedia.org/wiki/William_Maclure). Though Maclure displayed his map in Europe, it did not have the impact on continental geologists of the Brongniart-Cuvier map, whether because of his lack of scholarly connections or because his map and analysis were relatively simple, even backward, from a European perspective.

The map of the Paris Basin may have been inspired by a more ambitious one already started by the English mineral surveyor William Smith (1769–1839). Brongniart had seen an early draft of it when visiting London in 1802. Smith, the orphaned son of a village blacksmith, lacking connections to learned society, and often financially stressed, conceived the map and worked on it slowly but steadily, walking much of England and Wales, mostly alone, noting outcrops and from these charting his magnificent display, "A Delineation of the Strata of England and Wales," finished in color in 1815. Writer Simon Winchester's 2001 biography of Smith, *The Map that Changed the World*, is hyperbolic, but surely Smith's map was influential, and beautiful – and huge! It measures eight feet high and six feet wide (2.4 × 1.8 m); the various strata are colored in blue, green, yellow, red, and umber; and it now hangs above a grand marble staircase in London's Burlington House mansion. Images on the web are suggestive but do not do it justice.

Leaving aside the beauty of the thing, Smith went further than Brongniart and Cuvier in associating characteristic fossils with each stratum. When identifying fossil-containing layers, his watchword was to depend on the fossils more than on the composition of the rocks. If there were two exposures a considerable distance apart, each containing repeated alternations of, say, sandstone and limestone, then rocks alone could not connect a particular layer of limestone at one site with the corresponding limestone at the other site. But by looking for characteristic fossils in each layer, a distant connection could be made. *The fingerprint of each stratum was its fossil assemblage, not its rock composition.* This enduring principle, though not universally applicable, brought clarity to what had been ambiguous correlations of stratigraphic sequences in different regions.

Although Smith was not centrally interested in Earth's history, his association of particular fossils, or assemblages of fossils, with particular strata fortified the emerging interpretation during the early nineteenth century of fossils as stony clocks, timing the deposition of different sedimentary strata. The 1837 edition of British geologist Charles Lyell's *Principles of Geology*, by then a standard source (an earlier edition was carried by Darwin during his voyage on the Beagle), takes as given that Earth had a long history before the appearance of life, that its history is recorded in the successive strata and fossils, and that humans were late to appear.

In his 1841 catalogue of fossils from southern England, John Phillips, a nephew of William Smith, described three great eras of life with names coming into use then and still used today: Paleozoic, Mesozoic, and Cenozoic. He distinguished these by their distributions of fossils. The oldest, the Paleozoic, roughly comprised what had been called upper Primary and Secondary strata, containing fossils considerably different from animals still alive. The Mesozoic or middle era had plentiful ammonites (and dinosaurs). The newest era, the Cenozoic, has fossils identical or very similar to those known alive (Figure 2.5).

Furthermore, by counting species, genera, and families in the strata of each era, Phillips concluded that in each one, the number of kinds first increased, then greatly decreased, forming the distinct breaks between eras that are still recognized today.

Modern paleontologists and writers see the history of life punctuated by five "great extinctions," often with the cautionary note that we humans are currently causing the sixth (Leakey and Lewin 1996; Kolbert 2014). Best known of these is the one killing the last of the (nonavian) dinosaurs and many other animals, their coup de grâce widely accepted to have been a large asteroid or meteor impacting offshore today's town of Chicxulub on Mexico's Yucatan Peninsula, its crater identified. (Massive volcanic activity in central India may have independently contributed.) Modern dating puts the impact at 66 million years ago, marking the boundary between the Mesozoic and Cenozoic eras, which John Phillips noted though of course had no idea

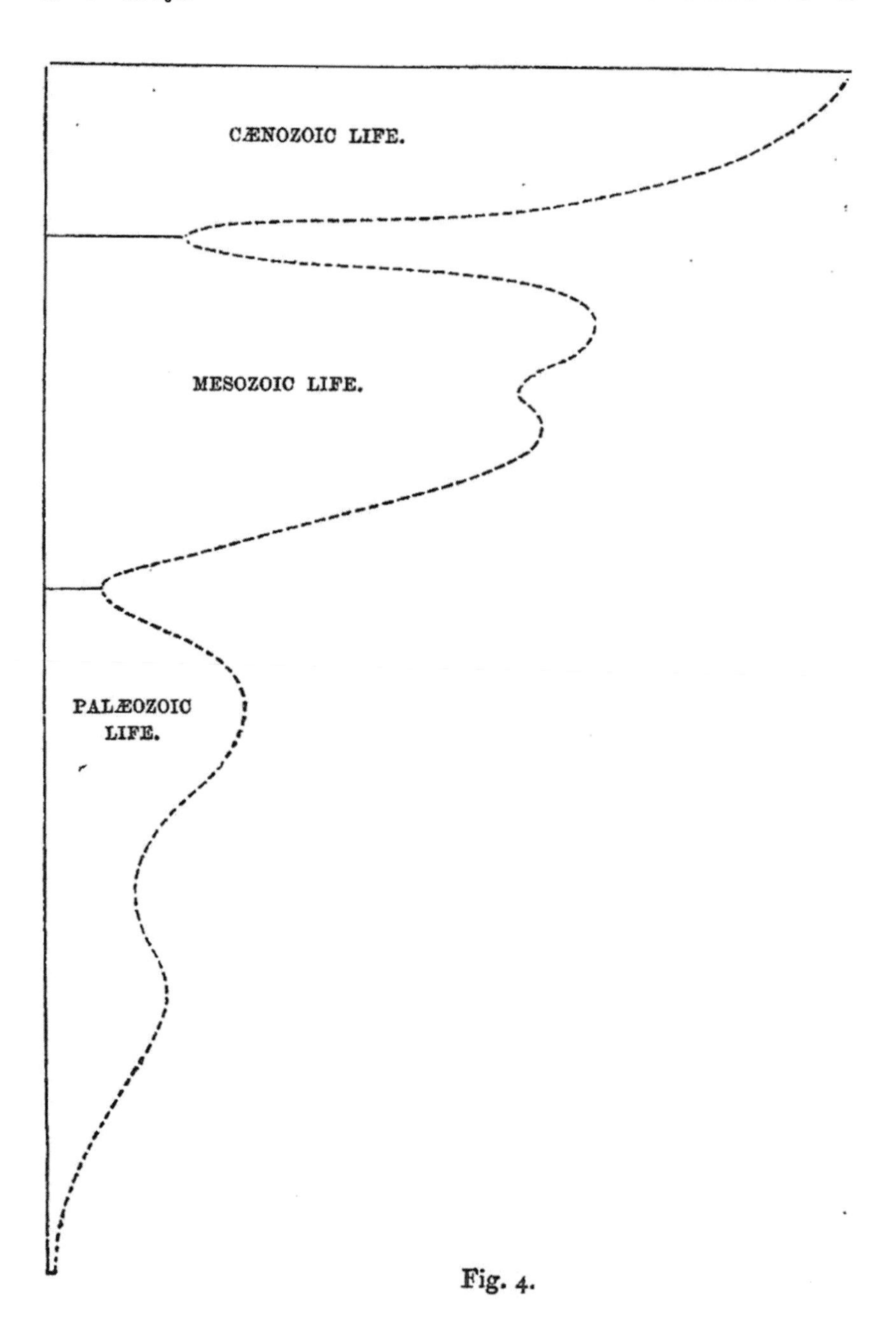

Figure 2.5 John Phillips's diagram of the three great eras of life.
Source: Phillips 1860: 66

how much time had since elapsed or what its cause. This was a great extinction but not the biggest.

The greatest of all extinctions, separating the Paleozoic from the Mesozoic in Phillips's diagram, is today called the "end Permian extinction," estimated to have wiped out 95 percent of marine and 70 percent of terrestrial species. (*Permian* is the name given to the youngest period – the top stratum – of Paleozoic rocks, named for the Russian city of Perm, where it was first demarcated.) Its cause is uncertain but suspected to be extensive and lengthy extrusion of subsurface magna, adding carbon dioxide to the atmosphere and acidifying the oceans.

Phillips was comparing *aggregates* of species, even as it was recognized that specific fossils could be used to identify specific strata, or at least groups of strata. Ammonites, for example, appear only in Mesozoic strata, not earlier or later. Furthermore, many kinds of ammonites were known, their appearance changing quickly from one stratum to another, and if widespread over regions, a particular kind of ammonite became – and still is – a reliable marker for an "horizon" in the stratigraphic pile. The same horizon in different regions could be connected if they contained the same kind of ammonite.

American Research

Whereas Europeans had so far dominated geology and paleontology, the United States soon became a center of research. This was a consequence largely of state governments forming agencies to explore their mineral wealth. Foremost was the New York State Geological and Natural History Survey, founded in 1836 to examine the state's stratigraphy and fossil history. By 1879, the federal government followed suit, creating the US Geological Survey, charged with exploring the geology and mineral resources of public lands, especially the vast areas added to the United States by the Louisiana Purchase in 1803 and the Mexican-American War in 1848.

Better exposures than in Europe marked the beginning and end of the Paleozoic by discontinuities in strata and the fossils they contain:

> [The Paleozoic] record is most completely preserved in the rocks of North America, and it is primarily by the labors of geologists in this field that this history has been uncovered ... On the

> whole, the organisms of the Paleozoic were peculiar to it and their history constituted a closed or almost closed chapter in the life record of the earth ... The change in the character of the life [has never been] as pronounced at any time as it was at the end of the Paleozoic. True, whole dynasties, such as those of the dinosaurs, of the ammonoids, etc., arose in the Mesozoic and disappeared at its close with surprising abruptness. Also, the ancestors of many Mesozoic forms are found in the later Paleozoic. Nevertheless, the change from Mesozoic to Cenozoic time is, on the whole, less abrupt than that which took place at the end of the Paleozoic. (Grabau 1921: 174)

The Cambrian, the earliest period of the Paleozoic era, was thought to contain the oldest fossils visible to the naked eye. This is not true. The oldest visible fossils are stromatolites, which are microbial communities packed together in layers, presenting a tangible mass. Stromatolites still live in shallow waters of Australia. While they are certainly visible, they comprise unicellular organisms that individually are microscopic and therefore their aggregates are not usually counted as visible creatures.

The enigmatic Ediacaran fossils are a more important exception, including many macroscopic but strange soft-bodied organisms that have evaded taxonomic classification. Their body plans are unlike and therefore difficult to compare with later phyla. Guesses about their phylogenetic placement include bacterial colonies, marine fungi, lichens, giant protists, and marine animals, while some paleontologists acknowledge having no idea what they are. Certainly, they were living prior to the Cambrian but are quite unlike Cambrian creatures. Ediacaran fossils were first reported in the 1940s from pre-Cambrian sediments in Australia. Recent analysis suggests they qualify as animals (Bobrovskiy et al. 2018). Perhaps they were an evolutionary dead end, leaving no modern descendants. I admit joining many writers who, after noting these pre-Cambrian anomalies, ignore their place in the greater scheme of things.

Typically, there is a marked boundary between pre-Cambrian rocks (without fossils) and Cambrian strata, sometimes termed the "Great Unconformity." In North America, and often elsewhere, the boundary separates underlying crystalline basement rock from much younger Cambrian sedimentary deposits (Peters and Gaines 2012).

This unconformity must have been caused by a lengthy period of erosion that smoothed the pre-Cambrian pile before Cambrian strata settled on top of it., with fossils more or less recognizable as known phyla (but no land plants or vertebrates). They appear so suddenly in diverse, profuse, and complex forms that their debut is often called "the Cambrian Explosion," as if living things, at least visible life, first began with a bang. It is difficult to say if the "explosion" truly reflects the introduction of modern phyla. Strata above the pre-Cambrian that have since eroded away, may have contained transitional forms of life showing a more gradual evolution from unicellular organisms to complex creatures. We may never know.

The Cambrian Period, the earliest to contain visible life, is named for the Latinized Welsh name for Wales, where the type exposures were first recognized, but the best exposures were found in North America (and much later in China). In 1909, the American paleontologist Charles Walcott of the Smithsonian Institution (1850–1927), working in Canada, discovered exquisitely preserved Cambrian fossils in the Burgess Shale of British Columbia. His fossil shells showed inordinately fine detail, and even soft parts, rarely preserved, were visible in the Burgess. Walcott returned in successive years with his wife and children, eventually collecting over 65,000 specimens, but he had no time to properly describe them, and their significance was not appreciated.

Beginning in the 1960s, the Burgess Shale was examined far more systematically by paleontologists from the University of Cambridge. These finds became famous when described in a popular book, *Wonderful Life: The Burgess Shale and the Nature of History* (1989), by celebrity paleontologist Stephen Jay Gould (1941–2002), but also controversial when some of Gould's claims were challenged by Simon Conway Morris (1998), who had done much of the excavation. (Gould had not dug there, depending on published reports and his vivid imagination.)

Modern dating puts the Cambrian Period at 541 to 485 million years ago, a duration of 56 million years. Most famous of its creatures are the trilobites, which were not only visible and shelled but surprisingly complex creatures for appearing so suddenly in the rocks, even with complex crystalline eyes. It was as if they were brought to Earth from elsewhere, already fully formed. They continued to evolve throughout the Paleozoic (see Figure 2.6 for post-Cambrian specimens) but did not outlast it. Some 20,000 species of trilobites are known, some tiny, some huge, some plain, some ornamentally spiny (Fortey 2000).

Figure 2.6 The trilobite *Phacops rana*, collected by the author and Matthew Tribe near Moravia, NY

The profuse variety of ancient trilobites is often likened to modern beetles, of which 300,000 species are known. There is a story circulated among entomologists that a theologian once asked the famous British biologist J. B. S. Haldane (1892–1964) what he had concluded about the nature of the Creator from his study of evolution. Haldane responded, "He had an inordinate fondness for beetles." Trilobites, too, at least until the great extinction ending the Paleozoic era.

Dinosaurs

Victorian anatomist Richard Owen first described the features of a taxon he named "dinosaur" based on fossils from three sites in England, but England is not a good source of dinosaur remains. Very soon, the western United States, its highly eroded landscape rich in fossils, became the treasure ground for dinosaur hunters.

Ignoring numerous technical distinctions that set dinosaurs apart from other reptiles, we can think simplistically that the earliest walking tetrapods – four-limbed creatures – had splayed limbs, extending outward from the torso. Most reptiles have splayed limbs, which, when walking, leave tracks in which the left and right prints are wide apart. Dinosaurs had limbs directly underneath their torsos, leaving tracks in which the left and right prints are closer together, a difference that apparently had advantages because dinosaurs became the dominant form of land life during the Mesozoic era, lasting 150 million years. Many dinosaurs have been identified from their characteristic footprints, which are fossilized more frequently than bones.

After the Civil War, two American paleontologists, Edward Drinker Cope (1840–1897) and Othniel Charles Marsh (1831–1899), personally and employing teams of workers, extracted fossils copiously from the arid West, vastly increasing our knowledge of the Age of Dinosaurs, all the while feuding acrimoniously (Jaffe 2001; Lanham 2011). For three decades, they famously and bitterly competed to see who could dig up the most spectacular finds, a rivalry known as "the Bone War." It was nasty and ungentlemanly, each man trying to get hold of the other's fossils and to undermine his reputation, costly to both in health and finances. But it was also productive. Between them, most of the giant species known to schoolchildren – *Triceratops*, *Allosaurus*, *Stegosaurus*, *Apatosaurus* (formerly *Brontosaurus*) – were discovered and sent for display at eastern museums. Some dinosaurs were small, but others were among the biggest creatures to walk the earth. (Living blue whales are the largest animal ever.)

Edward Cope came from a prosperous Quaker family in Philadelphia, free to pursue his avid interest in natural history, working part-time at the city's Academy of Natural Sciences, where he would spend his career. Largely self-taught, he never earned a university degree, nevertheless publishing his first scientific article while a teenager, eventually following it with a thousand more.

Othniel Marsh was older, from more modest beginnings, his father a farmer in rural New York, but he benefitted from a millionaire uncle, the banker George Peabody. With Peabody's support, Marsh graduated from Yale University, continued his education at Berlin and Heidelberg, and on returning to the US was appointed professor of vertebrate paleontology at Yale, the first professor of paleontology in the United States. The same year, the Peabody Museum of Natural History was founded at Yale with a donation from his uncle. In later life, Marsh was president of the National Academy of Sciences.

The young men first met in Berlin on friendly terms, staying together and maintaining correspondence. Later Cope hosted Marsh's visit to a New Jersey quarry where a dinosaur had been unearthed. Behind Cope's back, Marsh made an agreement with the quarry owners to deliver any new fossils to him. The relationship worsened after Cope published too quickly his drawing of a new species of meters-long reptile, inferred from its fossils. When Marsh pointed out that the skull was put at the end of the tail, not the neck, Cope's humiliation sealed their animosity.

In a competitive rush to claim new species, they sometimes coined extraneous names for different bones that later turned out to belong to a single species. Thus, Marsh in 1879 named a huge, nearly complete skeleton *Brontosaurus* ("thunder lizard"), which he thought to be an entirely new genus and species of sauropod dinosaur. In 1905, the American Museum of Natural History (AMNH) in New York City, using its own bones (with missing pieces fabricated), displayed the first-ever mounted skeleton. Brontosaurs thus became enshrined in the grammar school pantheon of hugely impressive dinosaurs. But its fossils were later identified with *Apatosaurus*, whose partial remains had been discovered and named earlier. By rules of classification, the earlier designation had priority. My childhood's beloved name "brontosaur" became obsolete, as serious a blow as when Pluto was demoted from planetary status.

Marsh shipped all his fossils to Peabody Museum at Yale, and soon other prominent museums became centers for paleontology: the Carnegie in Pittsburgh, and the Field Museum in Chicago, which I would visit as a boy, awed by its majestic displays of exotic and extinct animals. New York's AMNH in Central Park developed one of the best collections and had the biggest audience.

Under its aristocratic director, Henry Fairfield Osborn (1857–1935), the AMNH was especially active in collecting dinosaur and mammal fossils. Osborn was not above going into the field himself, but it was his employee Barnum Brown (1873–1963) who did the yeoman's work of finding and digging out, or trading for remarkable fossils, searching in the American West, in South America, and in Asia. Born on a Kansas farm, he was named after the showman P. T. Barnum and became himself a larger-than-life character, a dashing romantic hero, even a spy during World War II. His celebrity has by now dimmed: "In the mid-twentieth century, however, Brown – known by his adoring admirers as 'Mr. Bones' – was one of the most famous scientists in the world . . ., then as today . . . recognized as the greatest dinosaur collector of all time" (Dingus and Norell 2010: xi).

Of all his remarkable finds, the *pièce de résistance* was the type specimen of *Tyrannosaurus rex*. Discovered in the daunting Hell Creek Formation of Montana, embedded in hard matrix, its bones were excavated over the period 1902–1905. Six years later, Brown found another *T. rex*, also in Montana, a nearly perfect specimen, put on display at the American Museum. It was stunning, history's most fearsome carnivore, the stuff of nightmares.

In 1941, the AMNH sold the first specimen to Carnegie Museum at cost ($7,000) for fear that German airships might bomb the New York museum and destroy the second *Tyrannosaurus rex* skeleton mounted there. At least one specimen would be preserved. Since World War II, about forty more skeletons have been discovered. One, a juvenile collected on private land, was legally offered for sale on eBay, asking price $3 million (Telford and Shaban 2019). In 2020, an unusually complete skeleton discovered by an amateur was sold at auction for $31.8 million (Brainard 2020).

T. rex was found in upper Cretaceous rocks, so Brown knew it lived during the final act of the Age of Dinosaurs, close to the boundary John Phillips had noted between the Mesozoic and Cenozoic eras. Surely that was longer ago than the 6,000 years allotted by the Bible, but Brown had no more idea than Phillips how many thousands of years longer, or how the reign of *T. rex* ended, or what had happened in the meantime.

Today paleontologists recognize a boundary in the rock pile below which dinosaurs are found and above which they are absent. In some places this transition is extraordinarily sharp, appearing in the rocks near Gubbio, Italy as a centimeter-thick layer of clay. At Gubbio and elsewhere, the transition layer is unusually rich in iridium, an element rare in Earth's surface but abundant in asteroids, giving rise to the controversial theory that the impact of a large asteroid or meteor had abruptly ended the Age of Dinosaurs. An impact crater of the proper size and age was identified on the Yucatan Peninsula of Mexico, near the resort city of Cancun. Evidence grows that this indeed ended the reign of (nonavian) dinosaurs (Hull et al. 2020), marking the boundary between the Mesozoic and Cenozoic eras.

3 DARWIN'S REVOLUTION*

When savants of the early nineteenth century "burst the limits of time," they expanded Earth's history sufficiently that Louis Agassiz could imagine a very long age of icy climate (focus of the next chapter) and also for Darwin to conceive the slow modification and proliferation of creatures. Charles Darwin (1809–1882) described natural selection as a mechanism whereby this could happen and marshalled considerable evidence that it had happened. In his view, extinctions of species occurred gradually and continuously, usually because of the inability of the lost species to compete with another on the scene. If the fossil record seemed to indicate a sudden loss of many species, such as the trilobites or the dinosaurs, he thought it an illusion due to a gap in the stratigraphic record. Like his geologist friend Charles Lyell, Darwin dismissed the idea that extinctions were caused by great catastrophes (Raup 1995).

Within a decade of the 1859 publication of *On the Origin of Species*, Darwin's views were widely accepted among British literati. Publications and private correspondence of 67 British scientists, written between 1859 and 1869, show three-quarters accepting Darwinism, at least in the sense that each species evolved from another (Hull et al. 1978). Acceptance was similar in much of Europe, but less in the United States.

Historians of biology have long recognized that evolution was an idea whose time had come. It was "in the air," as they would say

* Portions of this chapter are adapted from A. Mazur, *A Romance in Natural History*, 2004.

(Eisley 1961). Welsh naturalist Alfred Russel Wallace (1823–1913) independently came up with the idea of natural selection while he was working in the Malay Archipelago. Darwin, alarmed on receiving a letter from Wallace outlining his own theory, arranged for both their ideas to be published simultaneously in 1858 (Desmond and Moore 1991). Wallace's letter spurred the long-ruminating Darwin to finish his great book the next year.

On the Origin of Species was released to a well-primed audience. Lamarck had taught the transmutation of species decades earlier, as had others. Even Darwin's grandfather, Erasmus Darwin, had published ideas akin to transmutation via survival of the fittest. Just 15 years before *Origin*, an anonymously written book, *Vestiges of the Natural History of Creation*, argued that everything currently in existence had developed from earlier forms, including the solar system, rocks, animals, and ultimately man. *Vestiges* was a publishing sensation, running through 14 editions, selling nearly 40,000 copies from 1859 to the end of the century, more than *Origin*. In the 12th edition, Scottish journalist Robert Chambers admitted he was the author (Secord 2000).

What is striking about the early history of evolutionary thought is how different the theories of transmutation were. Darwin himself believed that evolution occurs gradually, that operative changes are very small, with species emerging over many generations, without direction or progressive improvement. He (and Wallace) promoted one particular mechanism, natural selection, or "survival of the fittest," in Herbert Spencer's phrase. The basic idea is that a variety of traits always appears among members of a species, and that individuals with certain of these traits will be more successful in having healthy offspring. Therefore, those advantageous traits will be passed down, becoming more frequent in the next generation, and so on. Just which traits are advantageous will depend on the environment at that time (e.g., a cold or warm climate, kind of food available) and conditions for mating. These conditions change, favoring different traits at different times, so any appearance of progression or intelligent design is illusory.

In Germany, Ernst Haeckel (1834–1919), biologist and artist, became Darwin's chief proponent and popularizer, with his own twist. He was the first to draw an evolutionary "tree," showing the lowest life forms ascending progressively to Man (Figure 3.1). Like many early proponents, Haeckel believed that evolution was directional, that

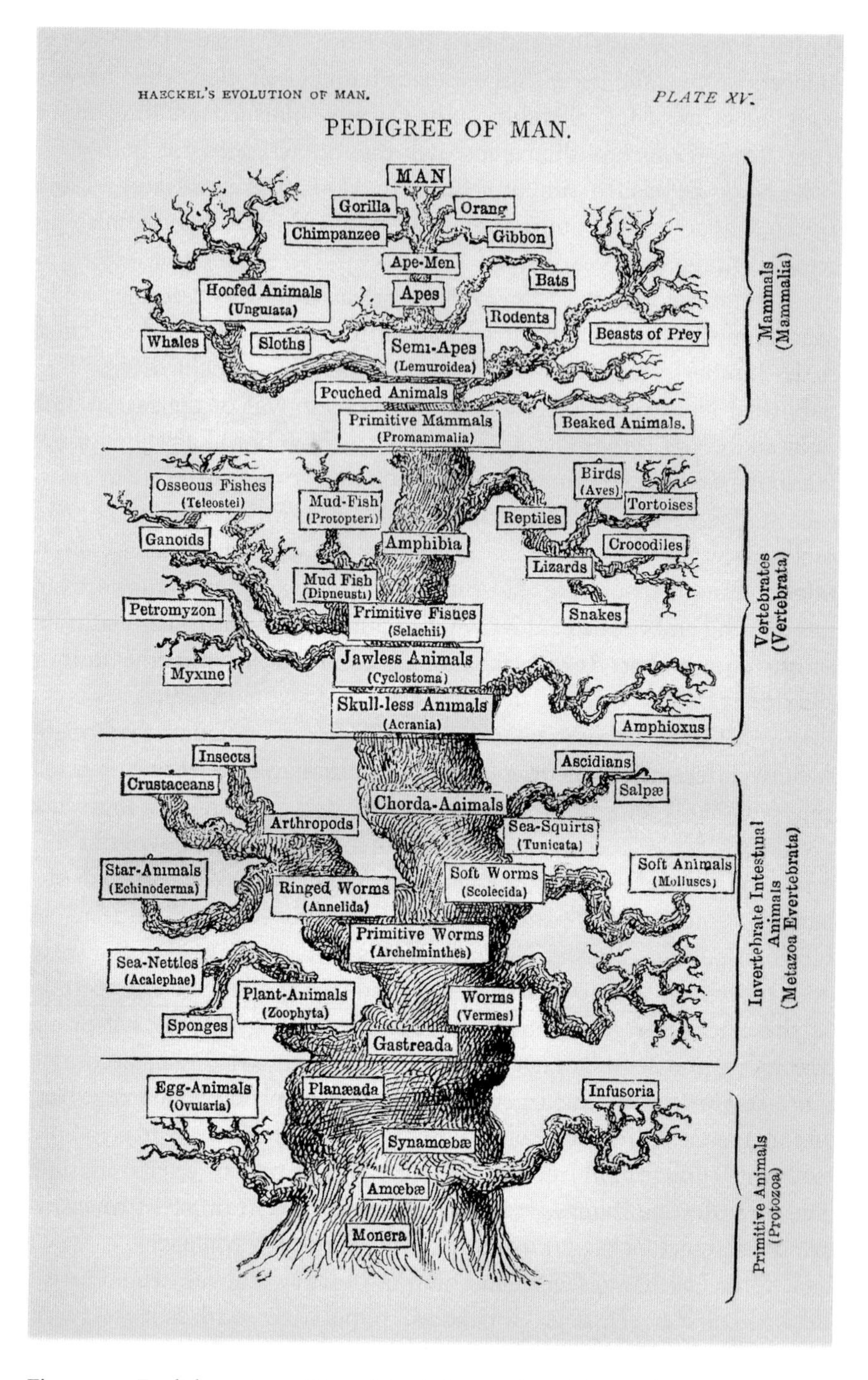

Figure 3.1 English version of Ernst Haeckel's tree from *The Evolution of Man* (1879). "Man" is at the crown of the tree. For Haeckel, as for many early evolutionists, evolution was progressive, reaching its pinnacle with European humans.

human beings ascended from primordial slime to become the superior creatures we are.

Haeckel also proposed the famous but now discredited law that "ontogeny recapitulates phylogeny," i.e., the developing individual passes through the same progressive changes that produced its species. Comparing developing embryos of living species, Haeckel claimed that an individual embryo resembles in turn its evolutionary ancestors. The human embryo, for example, looks first like the embryo of a fish, then an amphibian, then a mammal, and so on. There is a grain of truth here – human embryos do have gill slits at one point and further on look like monkey embryos – but Haeckel exaggerated these stages with the unintended result that many scientists became sceptical about comparing living species for our understanding of evolutionary change.

Apart from progressivism, the view of evolution most popular among scientists of the late nineteenth century, even those calling themselves "Darwinian," was saltation, that is, species change from one to another in discreet jumps. (One could hold saltationist views without progressivism, and vice versa.) Even the vocal proponent Thomas Huxley, "Darwin's bulldog," opted for evolution by saltation, with new species emerging in a generation. The notion of quick transmutation from one species to another (without progressive direction) continues today in the view called "punctuated equilibrium" (Eldredge and Gould 1972), but gradual transition is more commonly assumed.

In America, natural scientists were quickly drawn to one version or another of evolution, though Professor Louis Agassiz, "discoverer" of the Ice Age, remained a staunch opponent. His Harvard colleague, Professor Asa Gray, became Darwin's strongest supporter in the United States but believed in progressive evolution, that newer species are intrinsically superior to older species.

Alpheus Hyatt (1838–1902), a student of Agassiz, would become one of the leading American evolutionists while disagreeing with Darwin in important ways. The essence of natural selection is that the environment *selects* which kind of inherited trait will thrive and which kind will be winnowed. Darwin explicitly compared Nature's function with that of the human breeder of fancy pigeons, who selects a desired color or plumage to produce a specialty breed of show birds. Today, Darwin's mechanism (modified to incorporate modern genetics) is widely regarded as a correct explanation for speciation. But in Darwin's own day, and for decades after, it was hard for many committed evolutionists to swallow the notion that "blind" nature could passively produce such exquisite organs as the eye and brain.

Furthermore, doubters thought the shaping of species by an ever-changing and chaotic environment implied an equally chaotic or zigzag line of development. If the predominance of tiny, hard seeds encourages the profusion of birds with small, sharp beaks, then a change in climate that altered the prevailing seeds should push beak shape in another direction, and so on. Yet the fossil record showed many examples of continuous transitions, or lineages, among succeeding species. Indeed, the entire history of life, from Paleozoic to Cenozoic, suggested a generally "upward" progression from simple invertebrate forms though reptiles and mammals to primates and eventually humans. Paleontologists like Hyatt, while rejecting any divine or mystical guidance to the process, nonetheless sought in natural causes some explanation for the seemingly "straight line" evolution – later called "orthogenesis" – of living forms (Hyatt 1897a; Gould 2002).

Hyatt specialized in the study of ammonite mollusks, which evolved so rapidly that particular ammonite species were used as index fossils to distinguish subdivisions of the Mesozoic Era (Figure 3.2). If we were to watch the growth of an individual mollusk, we would see that it enlarges its shell in stages without losing or altering older parts. A mollusk embryo typically has a tiny, smooth, rounded shell. As it grows, it might add ribbing; then small protrusions; then in its adult stage it might add keeled whorls; in old age it may produce some unornamented surface. Thus, the shell of a dead mollusk shows a complete record of its owner's development while it lived. This is a paleontologist's delight, for with a single well-preserved adult fossil, one can trace the typical development of any member of that species. To make a similar reconstruction of a vertebrate's development, one would need separate fossils of individuals of that species in all stages of life, from embryo to old age.

Comparing fossils of advanced ammonite species with fossils of primitive species, Hyatt made a remarkable observation. The advanced shells contain all the features of the primitive shells, plus more. Hyatt concluded that individuals of the advanced species matured through all the stages of the primitive ammonites and then matured further, adding unique features. Note the similarity to Haeckel's law of recapitulation. (The Haeckel-Hyatt law of recapitulation was anticipated by Robert Chambers in *Vestiges*.)

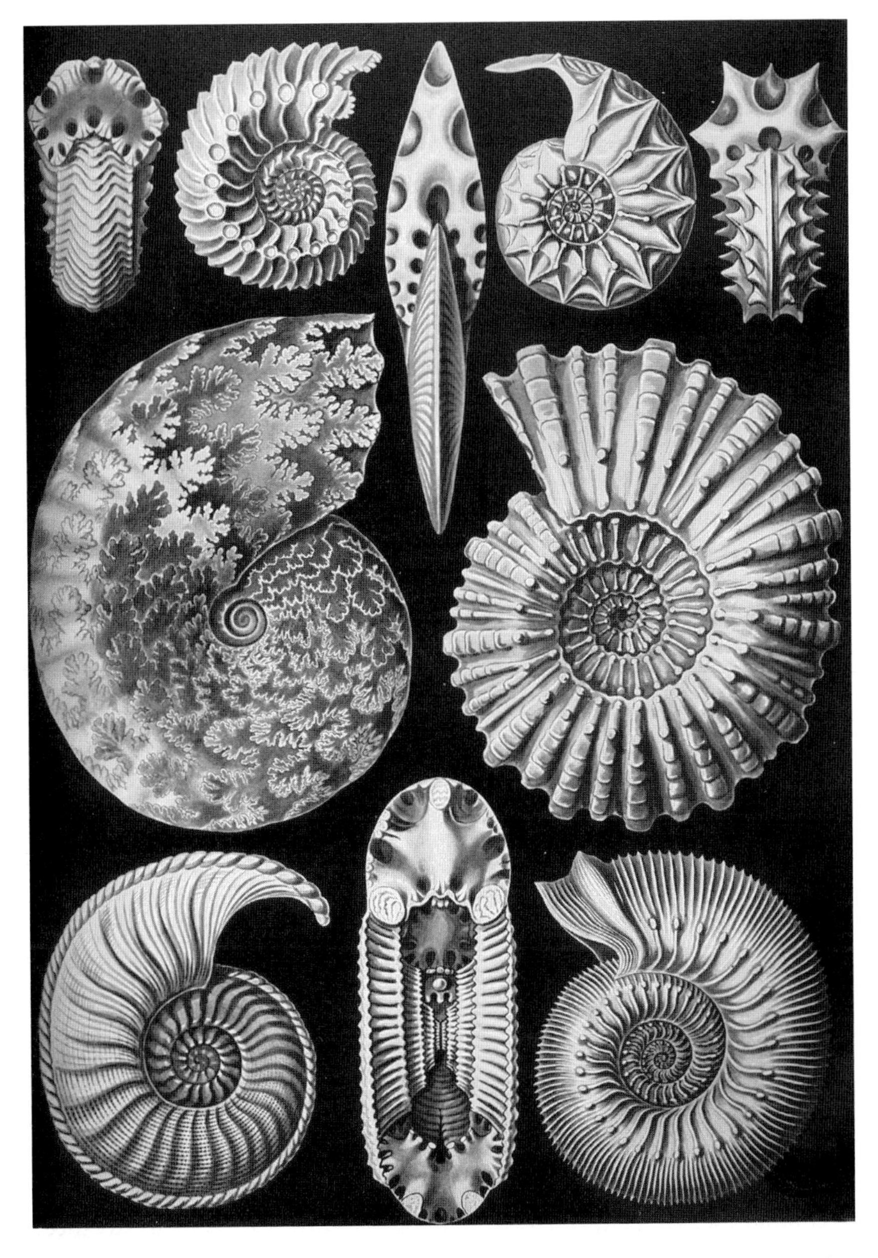

Figure 3.2 A variety of ammonite forms, drawn by Ernst Haeckel. Like most of Haeckel's illustrations, this one is as much artistry as biology.
Source: Haeckel, 1904, *Art Forms in Nature*

Hyatt thought the phylogeny of any lineage exactly reflects the ontogeny of individuals in its most advanced species. Just as individuals passed through fixed stages from birth to death, so the lineage of successive species passes through the same stages, from inception to extinction. We accept today that each fertilized egg contains genetic instructions for the reliable assembling of an adult. Hyatt thought entire evolutionary lineages follow instructions in their constituent organisms. If the ontogeny is smoothly directional, then the phylogeny is smoothly directional.

He went further. Ontogeny tells us not only the evolutionary past but also the evolutionary future. Since any adult eventually matures beyond its peak vitality, descending into senility, Hyatt thought lineages must do this too. He would arrange fossil adults from different ammonite species in sequences representing what he supposed was an evolutionary succession. The earliest species he regarded as youthful and vigorous, the later species progressing toward an acme of development, with still later species in decline. These later "gerontic" species had evolved too far, into maladaptive senility. All lineages would eventually have the same fate.

Darwin stressed the nondirectional shaping force of the external environment, while Hyatt believed the succession of species follows the internal instructions that guide the development of each organism. Darwin regarded the future of any lineage as uncertain, depending on environmental contingencies, while Hyatt claimed that each lineage's maturation and senescence are preordained in the same way that an individual's development is predictable. Darwin claimed that all evolutionary modifications are adaptive, while Hyatt asserted that modifications in a senile lineage are maladaptive. Irish elk died out because their antlers enlarged so much that they became unwieldy; sabertoothed cats went extinct because their lengthening canines interfered with eating. Human lineages would also pass their point of maximum vigor, become senile, then extinguishing. Darwin did not see much sense in Hyatt's orthogenesis, but in turn-of-the century America, Hyatt was the more influential voice, so straight-line evolution in a progressive direction was commonly assumed. It was the basis upon which human races were ranked, from primitive black Africans, through Asians, to superior white Europeans.

Much of the apparent continuity in the fossil record existed in the eye of the paleontologist; it was not truly there. The supposed evolution of the modern horse (*Equus*) is a classic example of superimposed linearity (Simpson 1951). For years, textbooks illustrated the seemingly straight-line (orthogenetic) transition from the small multi-toed *Hyracotherim* to today's large one-toed *Equus* as a smooth trend. With new fossil discoveries filling in the equine family tree, this straight-line scheme was shown to be incorrect. The tree has numerous additional branches going in different directions, all becoming extinct except for *Equus*. Similarly, human evolution often was (still is) illustrated as a straight-line development. No doubt readers have seen pictorial representations of linear human development, from stooping ape to upright man confidently striding into the future.

No orthodox evolutionist takes orthogenesis seriously anymore or is astonished at the seemingly impossible evolution of so exquisite an organ as the light-focusing eye. It is today accepted that the environment sometimes applies unidirectional selection pressure over extended periods, as when favoring larger organisms, so there have been some long trends in which evolving species got bigger (dinosaurs, whales), but there are also cases of dwarfism, especially on island species. The implausibly "miraculous" evolution of eye no longer seemed so miraculous after recognizing that it has evolved independently in squid, some marine worms, jellyfish, snails, spider, and trilobites (Morris 2003).

Defining Species

Classification of species predates our understanding of evolution. Human languages may have always had labels for important types of animals, and nearly all cultures use broader categories like *fish*, *birds*, *reptiles*, and *mammals* to encompass similar creatures. The well-known scheme of Carolus Linnaeus (1707–1778) is an elaboration of this notion, grouping species on the basis of similarity. Linnaeus thought the orderly arrangement of the living world reflected God's plan of creation, which could be adequately described using seven levels of nested categories: kingdom, phylum, class, order, family, genus, and species (remembered as "Kings play chess on fancy glass tables."). He placed similar-looking species, say, lions and tigers, in the same genus (*Panthera*), calling lions *Panthera leo* and tigers *Panthera tigris*. At a higher level of abstraction, Linnaeus placed these big cats and domestic cats in the family Felidae (cats). Going up another step, he placed Felidae along with dogs, bears, and similar flesh eaters into the order

Carnivora (carnivores). Then these were included with all animals that suckle their young in the class Mammalia (mammals), and these were lumped with other animals have a notochord – an elemental spinal column – into the phylum Chordata (chordates). Finally, all animals (organisms consuming food to obtain energy) were placed into the kingdom Animalia, separating them from the kingdom of plants (which obtain energy from the sun via photosynthesis).

In all classification schemes, the species is the fundamental unit because it conforms most closely to our notion of a "kind" of animal. Species are based on observations of animal bodies – or remnants of bodies – whereas higher categories (genus, family, etc.) are abstractions derived from comparisons among species. Despite its fundamental importance, a species is difficult to define. Similar appearance is not always adequate. Dogs of different breeds, although dissimilar in looks and size, are members of a single species according to another defining criterion: that mating produces fertile offspring.

The criterion of breeding fertile offspring cannot be tested on extinct species so paleontologists revert to visual appearance or, recently, surviving DNA, sorting fossils into one species or another depending on evaluations of similarity. In the past, degrees of similarity were judged qualitatively; today paleontologists evaluate them statistically, combining numerous measurements on the fossils, but just when one has a new species is a judgment call. Quantitative methods remain laced with subjectivity, and controversy abounds. There are vehement differences between "lumpers" who assign most fossils to a relatively few taxa, and "splitters" who invent a new species every time they uncover another tooth.

There is also the problem of birds, bats, and cows. Birds and bats have wings; bats and cows produce milk. Which two are most similar? Linnaeus knew God's plan contains such puzzles and solved them pretty well, here noting the wings of birds and bats are constructed differently, one using feathers, the other a skin membrane; also the "finger bones" are differently arranged within bat and bird wings. On the other side, the milk mechanism of bats and cows is nearly the same. Linnaeus therefore placed bats and cows together as mammals, a grouping from which birds are excluded.

Darwin had little difficulty using Linnaeus's scheme of nested categories because it resembles a family free: Lions and tigers belong to the same genus because they have a recent common ancestor; lions, tigers, and domestic cats are in the same family because all three have a common ancestor somewhat further back. But certain features of the Linnaean scheme have become troublesome. While God might have arranged life on seven levels, evolution did not. To describe the nearly continuous gradations in phylogenies, systematists began inserting additional levels: subphyla (e.g., vertebrates), subclasses (e.g., egg-laying mammals), suborders (e.g., marine carnivores), superfamilies, subspecies, and so on. The cladistic school of systematics rejects this kludging, eschewing all Linnaean levels above genus and species. Traditionalists retain the oft-modified levels but rarely give them deep meaning.

Cladists further object to the Linnaean scheme because it classifies animals into groups that may not reflect lines of descent. Striving for consistency, cladists argue that any animal descended from, say, dinosaurs is itself a dinosaur. Since birds evolved from dinosaurs, birds should be regarded as dinosaurs, not placed in a separate class (Aves), as Linnaeus did. But this viewpoint has problems of its own. By the same logic, amphibians, reptiles, birds, and mammals (including humans) should all be classified as fish.

Our Current View of the History of Life

The post–World War II invention and improvement of radiocarbon dating enabled assignment of absolute times to fossils of plants and animals that lived as far back as 50,000 years ago, or Near Time. Subsequently developed dating techniques, both geological and astronomical, reach much further back in time, placing the age of Earth's crust, as well at the moon and visiting meteorites, at 4.5 billion years. The universe itself, beginning with the "big bang," is 13.8 billion years old, though no one knows what, if anything, existed prior to that point.

The Cambrian explosion, when complex invertebrates first appear, occurred 550 million years ago (mya). By 400 mya, bony fish, the first major group of vertebrates, had evolved from invertebrates. The great Permian extinction, finishing off the trilobites and ending the Paleozoic Era, was 250 mya.

Before 350 mya, the first tetrapods (four-limbed vertebrates, usually living on land) evolved from fish, perhaps from air-breathing lungfish or coelacanths whose lobed fins work like protolimbs. Subsequent tetrapods – mammals and birds as well as amphibians and reptiles – generally have the same skeletal plan. For example, the forelimb (our arm, a bird's wing) has one upper bone closest to the body, two bones below the elbow, several smaller bones in the wrist, and then longer finger bones. Forelimbs have disappeared in snakes but were present in their ancestors. Whales and birds have no external fingers, but "finger bones" are easily seen in their skeletons. All tetrapod skeletons, however modified, resemble the same ancestral structure.

The original land animals were amphibians (tetrapods that lay eggs in water), returning to the water to reproduce, their young hatching as aquatic larvae. Some of their descendants had, by 300 mya, developed "amniote" eggs with tough watertight coverings and large yolk for nourishing the embryo. Immersion in water was no longer necessary for reproduction. These amniotes, the protoreptiles, could live (and love) fully on land.

Systematists attribute great importance to the number of holes – one or two – in the cheek of an amniote skull. Though seemingly without functional significance, the number of holes marks an important evolutionary branch point. Early amniotes with two cheek holes – called "diapsids" – are ancestral to today's reptiles, while amniotes with one cheek hole – called "synapsids" – are ancestral to mammals. (Fish and amphibian cheeks have no holes.)

Reptiles, among the few species surviving the great Permian extinction, diversified into niches vacated by those less fortunate. Most spectacular were the dinosaurs, dominating the land from about 200 to 66 mya when the Yucatan asteroid struck. I have on my shelf two coprolites – fossilized dinosaur feces – abundant in the American West because of the long tenure and huge throughput of these voracious creatures.

Birds (warm-blooded diapsids with feathers) are descended from dinosaurs. Numerous fossils dating from 150 mya, including the famous *Archaeopteryx*, show feathers on creatures otherwise resembling small meat-eating dinosaurs. Paleontologists enjoy telling laypeople that dinosaurs are not extinct but live on as birds.

The first mammals (warm-blooded synapsids with hair and milk glands) predate birds, appearing about 220 mya. Small and inconspicuous at first, mammals would become beneficiaries of the dinosaur extinction, diversifying to fill vacant niches. Our own order of mammals, the primates, appear by 55 mya.

These numbers are too big for most of us to comprehend. It may help to think of the total span of Earth's existence as being compressed into one calendar year, its cooling into a solid sphere in January. The first microscopic life appears in March or April, but it was not until November that fish with backbones swam in the oceans. Dinosaurs flourished and then died off in the middle weeks of December, to be replaced at about Christmastime by mammals as the dominant land animals.

Some of these mammals were the early primates, whose main features were their grasping hands (rather than claws) and eyes in front of their heads (rather than the side) for stereoscopic vision, adaptations well suited for a life of climbing through trees. In the days between Christmas and New Year, some of the primitive primates evolved into monkeys (by December 27), and some monkeys evolved into early apes, the ancestors of present-day gorillas and chimpanzees (about December 29). Some of the early apes were our own ancestors, their descendants eventually becoming enough like us – more human than apelike – to call them *hominins*, and some of these became our own biological genus, *Homo*. This was about 2 million years ago or, in our compressed time scale, after supper on New Year's Eve, near the beginning of the Pleistocene Ice Age. Fully human members of our own species, *Homo sapiens*, appear by 10 minutes to 12, and with the invention of writing, all of recorded history fits into the minute before midnight.

Sketching this picture would have been impossible prior to the realization that Earth is far older than portrayed in the Bible, and that humans and our closest relatives are recent arrivals on the scene. Comparative anatomy and the discovery of ancient fossils, datable at least on a relative time scale, were essential methods for working out the history of life and relations among the various kinds of life. Darwinian evolution makes sense of the various transmutations seen in the fossil record and how the many kinds of life, both living and extinct, are related to one another.

By 2 million years ago, near the beginning of the Ice Age, there were primates in Africa whom we include in the genus of humans, *Homo*, though clearly different from us. By 200 thousand years ago there were people whose fossil skulls fit within the normal variability found among present-day populations. Some knew the Ice Age first hand, living by the margins of snow and ice that then covered much of the Northern Hemisphere, but cultural memories of that time were long gone by the nineteenth century. It required a wholly new discovery for the Ice Age to be recognized.

4 DISCOVERING AN AGE OF ICE

Louis Agassiz (1807–1873) is best known and admired as the man who discovered the Ice Age, but from a modern perspective, he does not look good. He opposed Darwin, never accepting evolution, asserting that species were fixed, and that whenever new forms of life appeared in the fossil record they were results of separate acts of divine creation. He deserted his first wife in Switzerland to accept a position at Harvard; he could be an arrogant tyrant to his students, rarely admitting error; and he vigorously advocated that "Negroes" were an inferior form of human. Even his recent biographer, Christoph Irmscher (2013), is alternately attracted to and repulsed by him. And Agassiz was not actually the first to discern that present day glaciers were once much greater in extent.

What he did do, and very well, was accumulate a convincing weight of evidence that an Ice Age had occurred, proselytized the idea, and won general acceptance for it. He was an intrepid field worker, a brilliant lecturer and teacher, a driving force in American science during the mid-nineteenth century, and an indefatigable worker. Charmingly convivial, he became a close friend of some of the major intellects of his time, including Georges Cuvier, Alexander von Humboldt, Henry Wadsworth Longfellow, Ralph Waldo Emerson, and Oliver Wendell Holmes.

Agassiz was born and grew up in the small Swiss village of Môtier, facing the Alps, with the Lake of Neuchâtel a short hike away. His father was a Calvinist minister, his mother the daughter of a physician living nearby. Louis was the first of his parents' children to survive infancy, followed by a brother and two sisters. A good student and avid hiker, he enjoyed the landscape around him, early on developing an interest in nature. At the age of seventeen, he was reading Lamarck and Cuvier, the most important biologists of his time. Financed by an uncle, he entered Heidelberg University and then the University of Munich to study medicine and natural history. Under the tutelage of a botany professor at Munich who had collected specimens in Brazil, Agassiz published his first book, on Amazonian fishes. The next year, at age 23, he completed two doctorates, one of them in medicine. Then he moved to Paris to study comparative anatomy under Georges Cuvier at the *Jardin des Plantes*, the foremost research institution in natural science, where Lamarck also worked. These two giants were in opposition, Lamarck teaching the transmutation of species, one into another, while Curvier was committed to the constancy of each species, from its first appearance to its extinction.

Agassiz was assigned to Cuvier's vast collection of fossil fish, and the two men developed a good relationship but it lasted less than a year, cut short by Cuvier's death. Cuvier's fossils were a gold mine for Louis, whose publications about them quickly established him as an authority on fossil ichthyology. Moving back to Switzerland, he accepted an appointment as professor of natural history at a small school in Neuchâtel and married Cecile Agassiz. They had three children, of whom the first, Alexander, would eventually succeed his father as a Harvard professor.

Sadly, the Agassiz marriage was failing, and perhaps teaching science to children was wearing. What sustained his enthusiasm, aside from fossil fish, were his trekking through Switzerland's mountains and a new interest in glaciers.

Erratic Boulders

A puzzle for early geologists was the presence of huge boulders where they had no business being, some as big as a house and a few of them famous for their size and anomalous placement (Figure 4.1). Called "erratics," those on a plain had no nearby mountain from which they could have rolled, and those perched high on a mountain had no higher point from which they might have descended. Erratics were often composed of a kind of rock that did not exist nearby but was known from a formation far way. Thus, they appeared to have been transported considerable distances from an original source to their present

Figure 4.1 An American example of an erratic boulder, Doane Rock is the largest glacial erratic on Cape Cod, MA. Pits dug at the base showed as much rock below the surface as above.
Source: US Geological Survey

locations, but how? There were smaller rocks, even pebbles, their composition also showing them out of place, but their means of transport was not so inconceivable, most likely by flowing water.

Well before Agassiz became interested in glaciers, several people living in Alpine regions had proposed them as the means by which these huge boulders moved from any plausible source. Historian Albert Carozzi (1967) awards priority to a Swiss peasant, the chamois-skin hunter Jean-Pierre Perraudin, who in 1815 told the savant Jean de Charpentier that Alpine glaciers once extended much farther than at present, were much larger, and moved boulders too big to have been transported by water. Furthermore, the moving glaciers produced striated and polished rocks, visible under present glaciers. De Charpentier at first did not believe Perraudin. But a few years later, his friend, the highway engineer Ignace Venetz, was more receptive to Perraudin's ideas, and in turn rekindled interest in De Charpentier and a few others. In 1834, De Charpentier presented a paper to the Swiss

Society of Natural Science, meeting in Lucerne, explaining his and Venetz's ideas about the transport of boulders by glaciers, and that during an earlier colder period the glaciers were larger in extent. His audience was unconvinced.

Untutored Swiss mountaineers accepted these ideas long before they became grist for scientific debate. De Charpentier relates that on his way to Lucerne to deliver his talk, he met a woodcutter who told him that a large boulder of Grimsel granite, lying by the road, was one of many stones in the area that came from far way, from the Grimsel, because they are composed of the same kind of granite, which is otherwise not found in the immediate vicinity. De Charpentier continues,

> When I asked him how he thought that these stones had reached their location, he answered without hesitation: "The Grimsel glacier transported and deposited them on both sides of the valley, because that glacier extended in the past as far as the town of Bern...This good old man would never have dreamed that I was carrying in my pocket a manuscript in favor of his hypothesis. He was greatly astonished when he saw how pleased I was by his geological explanation, and when I gave him some money to drink to the memory of the ancient Grimsel glacier." (quoted by Carozzi in his introduction to Agassiz 1967: xv)

The young Agassiz had met and liked Jean de Charpentier. In 1836, visiting the Alps during his summer vacation from school, Agassiz stayed with Charpentier. They, along with Venetz, visited some of the Alpine glaciers, and Agassiz became a convert to the glacial theory, whether from viewing the field evidence or because of the attention given him by his older and respected companions. Returning to Neuchâtel in November, Agassiz found polished rocks and erratic bounders in the nearby Jura Mountains. Joined by botanist Karl Friedrich Schimper, a friend from Munich, the two speculated about a past cold period with immense glaciers. It was Schimper who came up with the term *Eiszeit* (Ice Age).

In July 1837, at age 30, *wunderkind* Louis Agassiz gave the presidential speech to the annual meeting of the Swiss Society of Natural Science, held that year in Neuchâtel. The members, expecting their young president to discuss his specialty, fossil fish, were surprised to hear a talk on glacial theory. That ego-strong Agassiz was at this point a mere dabbler in this new area did not deter him. Combining much that he learned from De Charpentier and Venetz with his own explorations of local sites, he described his and Schimper's glacial theory. It was not well received. Carozzi (1967), who published an English translation of the paper, describes audience reaction as astonishment, incredulity, and commotion. The prominent Prussian geologist Leopold von Buch was "horrified," says Carozzi, his disagreement with Agassiz becoming violent and almost out of hand. Despite this dispute, Agassiz and the older von Buch remained friendly.

The next day Agassiz led a day trip to the Jura Mountains, joined by von Buch and another distinguished geologist, Elie de Beaumont, among others, to show evidence in the field. But he won few converts.

Considering that even Swiss peasants accepted that glaciers transported huge boulders, and that evidence of glacier advance and recession was plain to the eye, if one knew where to look, it is puzzling that so many objected vehemently when Agassiz presented the glacial theory in his presidential address. Why this resistance?

We can discount the claim sometimes offered that the theory was not compatible with Genesis. Rudwick (2008; 2014) convincingly shows that by the early nineteeth century, few geological savants, even those devoutly religious, were wedded to a literal interpretation of Bible stories about a Creation only 6,000 years earlier, or the appearance of all plants and animals within six days (of 24 hours), or of Noah's worldwide Deluge.

We can understand moderate objections even raised by some of Agassiz's friends and acquaintances. One, the prominent if eccentric Oxford geologist William Buckland, wrote, "I am sorry that I cannot entirely adopt the new theory you advocate to explain transported blocks by moraines; for supposing it adequate to explain the phenomena of Switzerland, it would not apply to the granite blocks and transported gravel of England, which I can only explain by referring to currents of water" (quoted in E. Agassiz 1885: Loc. 2416). Buckland was a "character" at Oxford who wore his academic gown and top hat into the field. Joining Agassiz in exploring Scotland, he was soon converted to the glacial theory.

There were already accepted explanations for transportation of erratic boulders, recapped by Charles Lyell in his influential *Principles of Geology* (1837). One was that they were swept along by rapidly flowing water, mud, or ice rafts in rivers, or carried on floating icebergs. Helping to explain the movement even of very heavy objects, there were known cases of bursting ice dams, suddenly releasing colossal flows of water. Lyell proposed that the encrustation of boulders in ice, reducing the specific gravity of the rock alone, would ease their transport. That flowing glaciers could be one vehicle for transport was itself uncontroversial. Indeed, boulders could be observed embedded in glacial ice and icebergs.

Another friendly critic was Alexander von Humboldt (1769–1859), famous for his explorations of South America. They had met while Agassiz was studying in Paris. The old naturalist took a liking to the talented young Swiss and his excellent comparative work on fish, becoming a quasi-father, correspondent, and adviser.

Writing in 1840, and referring to his earlier letters, von Humboldt apologizes for what may have been his too sharp objections to Louis's "vast geological conceptions," wishing that he would return to his far more worthwhile studies of fish. The old man continued, "Taught from my youth to believe that the organization of past times was somewhat tropical in character, and startled therefore at these glacial interruptions, I cried 'Heresy!'" (quoted in E. Agassiz 1885: Loc. 2626).

Von Humboldt was typical in believing that Europe's prehistoric climate was warmer, not colder. The fossil record showed animals and plants resembling those currently known to live in the tropics. This was consistent with belief that Earth has a hot core, deduced from rising temperature as one descends into deep mine shafts. Physics dictates that heat from this hot body is lost to space, implying that former temperatures were warmer than at present. Also, many of the critics, certainly Lyell, were committed uniformitarians, not highly receptive to the notion of dramatic changes in Earth's climate.

Such objections, moderate and coming from friendly critics, do not explain the vehement resistance to Agassiz's glacial theory. The reason for this intensity of opposition becomes clearer if we look at the startling claims at the end of his talk. Agassiz envisioned for his audience a sheet of ice extending from the North Pole at least to the shores of the Mediterranean and the Caspian Sea., despite the fact that no glacial deposits had been found in these southern areas. He spoke of a Siberian winter established on ground previously covered by a rich vegetation and inhabited by great mammals, similar to those now living in the warm regions of India and Africa, and of all this life being extinguished. "Death enveloped all nature in a shroud." These claims of cataclysmic events went far beyond the evidence at hand.

Perhaps most startling was his attempt to explain why temperatures during his putative Ice Age were colder than at present, whereas most geologists, like von Humboldt, saw evidence for a warmer past climate. Agassiz pointed out that different epochs are marked by the appearance of new species, which persist for a while, then go extinct, to be replaced by newer species in the next epoch. He then asserted an analogy to an individual organism, which begins life with the generation of heat, lives in a temperature equilibrium, and dies with the heat of life gone. He posited, with no apparent justification, that this held true for epochs too. The end of each epoch – the death of its species – is, like the death of an organism, accompanied by a loss of heat. During these temporary dips in temperature, an Ice Age could occur.

> The drop of temperature took place at the times of disappearance of (an epoch's) inhabitants. This drop reached a value below the temperature characteristic of the following epoch, which rose with the development of the newly appeared organisms ... There must have been a drop of temperature well below its present-day value, at the epoch which preceded the uplifting of the Alps and the appearance of present organisms. This drop of temperature must have been responsible for the formation of the huge masses of ice which covered the earth all over the places where erratic boulders are associated with polished rocks like ours. Undoubtedly, this great cold also buried the mammoths of Siberia in ice, froze all our lakes and accumulated ice up to the elevation of the summits of our Jura ...." (Agassiz, translated by Carozzi, 1967: liv)

This was egregious overreach, leaping from the regional case for the glacial theory to a global scale, and from the death of an organism to a cold spell at the end of each epoch. That he turned out to be correct about an earlier Ice Age should not detract from the implausibility of his claims at the time he first made them. Enlightenment geologists were disinclined to sit quietly for grand speculations unaccompanied by supporting evidence. Ironically, in later life as an ardent opponent of natural selection, Agassiz charged Darwin with grossly speculative theorizing.

The 1837 presidential address would become the core of a book, *Études sur les glaciers*, published three years later, after Agassiz and his colleagues had committed themselves to careful observation in the nearby Alps. He describes glacial movement, advance and retreat, and its tell-tale signs: moraines, polished rock, and striations etched by stones in the flowing ice. His most impressive proof of advancing glaciers is the displacement of a hut built by Franz Joseph Hugi in 1827 at the foot of the Aar glacier near a great block of granite. When Hugi returned in 1830, he found the hut pushed 100 feet (33 m) from the granite block, and six years after that he found it 2,200 feet (670 m) away. When Agassiz visited in 1839 he found the distance 4,400 feet (1340 m) from the great granite block, and by the next year it had moved another 200 feet (60 m). Perhaps the book's greatest impact came from the handsomely etched plates in an accompanying atlas, such as Figure 4.2, most illustrating the river-like appearance of actual glaciers (Carozzi 1967).

The book certainly made an impact, some of its accompanying plates reproduced even in North America, but caused bad feelings among two of Agassiz's close friends. It was arguably a breach of etiquette for Agassiz to publish his book just before one by the slower-writing de Charpentier, stealing thunder from the older man who had taught Agassiz much that he knew about glaciers. Schimper, Agassiz's coworker in developing the *Eiszeit* theory, also felt slighted. In all, he lost the friendship of more of his former allies than of his opponents.

The book did not dispel opposition to the Ice Age hypothesis because Agassiz persisted in his extravagant claims, going far beyond any supporting evidence. In 1865, while on an expedition to South America and seeing signs that Andean glaciers had once extended beyond their present limits, he jumped to the conclusion that European and North American ice sheets had extended into South America. His friend Lyell, who by that time accepted the northern ice sheets, wrote: "Agassiz ... has gone wild about glaciers ..., [claiming] the whole of the great [Amazon] valley, down to its mouth was filled by ice ... [Yet] he does not pretend to have met with a single glaciated pebble or polished rock" (quoted in lmbrie and Imbrie 1979: 45).

Acceptance

One reason that geologists rejected Agassiz's claims was that they had never seen or heard of ice formations on the scale that he

Figure 4.2 Glacier of Zermatt, Switzerland. Note the woman in midground for scale.
Source: Plate 4 accompanying *Études sur les glaciers* (Agassiz 1840)

envisioned. But by the mid-1800s, there were several reports by intrepid explorers of immense ice accumulation on Greenland and Antarctica.

From 1839 to 1843, the British admiralty had two warships with specially strengthened wooden hulls under the command of Captain James Clark Ross on a scientific exploration of the Antarctic, its main purpose to find the position of the South Magnet Pole. After reaching the highest southern latitude yet attained, Ross's attempts to penetrate further were frustrated by a precipitous wall of ice rising out of the water to a height of 150 feet (45 m) in places. He sailed 450 miles (724 km) in front of this cliff and found it unbroken by any inlet. This would later be known as the Ross Ice Shelf, the largest of Antarctica, and an extant image of Agassiz's vision.

Also by mid-century, there were voyages along the northerly coasts of Greenland in waters normally too ice-blocked and hazardous to sail. That huge island had been inhabited for four millennia by various Arctic people whose forebears migrated from what is now Canada or Alaska. Norsemen led by Erik the Red settled the uninhabited southwestern coastal area in the tenth century, and Inuit arrived in the thirteenth century. These populations waxed and waned with improvement or decline in Greenland's climate. The Norse disappeared altogether in the fifteenth century for reasons unknown. Those remaining, mostly Inuit, clustered on the southwestern coast, the most habitable part of the island. Elsewhere the coastal waters, and especially the interior, were inhospitable and barely known (Gertner 2019).

Voyagers up the coast of Greenland could see enormous ice sheets, far larger than Alpine glaciers, flowing into the sea (Figure 4.3). As ice flowed over the land's edge, huge icebergs calved off to float south. These ice sheets were "proofs of concept," helping to quiet opposing voices. Within another ten years, the glacial theory was generally accepted by European and American geologists, if not to the extremes to which its main author had pushed it (Krüger 2013). An advantage Agassiz had over other proponents of glacial action was that he was already famous for this work on fossil fish, and he had a network of prominent friends who eventually became supporters. William Buckland, initially doubting that glaciers transported granite blocks and gravel in England, soon decided they did and proselytized others, most importantly Lyell. Buckland announced the good news in a letter to Agassiz dated October 15, 1840:

Figure 4.3 Immense Greenland glacier, flowing into sea, by H. M. Skae. For scale, note the sailing ship near the ice face.
Source: Geikie 1874, Plate II

> Lyell has adopted your theory in toto!! On my showing him a beautiful cluster of moraines, within two miles of his father's house, he instantly accepted it, as solving a host of difficulties that have all his life embarrassed him. And not these only, but similar moraines and detritus of moraines, that cover half of the adjoining counties are explicable on your theory, and he has consented to my proposal that he should immediately lay them all down on a map of the county and describe then in a paper to be read the day after yours at the Geological Society. I propose to give in my adhesion by reading, the same day with yours, as a sequel to your paper, a list of localities where I have observed similar glacial detritus in Scotland, since I left you, and in various parts of England. (quoted by E. Agassiz, 1885: Loc. 2572)

With sponsorship from the King of Prussia, in whose domain Neuchâtel was then included, Agassiz turned to more careful empirical investigation, focused on the Aar glacier, its structure and the mechanics of movement. In one simple but ingenious experiment, he placed a series of stakes across the glacier, perpendicular to its direction of flow. Returning the next year, he found that stakes in the middle of the glacier had traveled farther than those at its sides. These and other empirical investigations were included in another book, *Nouvelles études et experiences sur les glaciers actuels* (1847), the first volume in a projected three-volume work, to be coauthored with his assistants, never completed because of Agassiz's departure from Europe.

After leaving this sequel with the printer in Paris, Agassiz traveled to the United States for what was intended as a temporary stay. He was delighted to find, almost as soon as he arrived, evidence for glaciation.

> In the autumn of 1846 ... I sailed for America. When the steamer stopped at Halifax, eager to set foot on the new continent so full of promise for me, I sprang on shore and started at a brisk pace for the heights above the landing. On the first undisturbed ground, after leaving the town, I was met by the familiar signs, the polished surfaces, the furrows and scratches, the line engraving so well known in the Old World; and I became convinced of what I had already anticipated as the logical sequence of my previous investigations, that here also this great agent had been at work. (quoted in E. Agassiz 1885: Loc. 3674)

This was not an especially lucky verification for his first exploration ashore. Northeastern America is awash in remnants of its last glaciation: erratic boulders, polished and striated bedrock, and moraines (Raymo and Raymo 2007). Geologists turned to their closer study, even examining the point of contact between a glacier and its bedrock, which could be observed directly in summertime when glaciers shrank.

> Creeping in below the ice, which it is often possible to do for some little distance, we find the rocks finely smoothed and polished, and showing long striae and ruts, that run parallel to the course followed by the glacier. If we pick out some of the stones that are sure to be scattered about below the ice we shall find that many are smoothed, polished, and striated in the same manner as the surface of the rock itself ... Could the glacier be removed, we should find the whole bottom of the valley smoothed and polished, and streaked with long parallel ruts. Every high projecting boss would be rounded and dressed on the side that looked up the valley; while the rock on the lee side, sheltered from the attack of ice-plough, and chisel, and graver, would retain all its roughnesses ... The finer-grained materials employed by the ice in polishing its bed, the impalpable mud and silt, are carried out beneath by the stream that issues at the foot of the glacier. In this manner, almost all glacier rivers have imparted to them a turbid appearance, the color of the water depending upon that of the sediment which it holds in suspension. (Geikie 1874: 45)

Figure 4.4 Marks of glaciation were abundant in America, as in Europe. Left: Glacial striations at Mt. Rainier National Park, WA. Right: Glacial polish at Yosemite National Park, CA, photo by Greg Stock. *Source*: U.S. National Park Service

Larger moraines, those chaotic jumbles of rock transported to the margins of a glacier or left at its terminus as the ice melted back, were mapped to show the farthest reach of the ice. In North America, a great terminal moraine formed a ridge, extending from eastern Long Island to Washington State. The finer sediments in moraines, the pebbles and silt abraded and carried by the flowing ice, emerged as important evidence of now-vanished ice sheets. Deposits of these unsorted sediments, insignificant in size, were commonly found north of the great American ridge moraine but not to the south, as if differentiating the former presence of an ice sheet from its absence. Called "glacial tills," they became an important if undramatic sign of glacial extent.

In perhaps the first book since Agassiz on evidence for an Ice Age, James Geikie (1874) devoted an entire opening chapter to tills, describing them as the lowest and therefore oldest of Scotland's superficial layers, tough tenacious clay crammed with a pell-mell assemblage of stones of all shapes and sizes, often showing smoothed, polished, or scratched faces. Such deposits were found mostly in lowland valleys, running between rounded hills. Scotland also had erratic boulders and U-shaped valleys, further marks of glacier flow. Altogether, Geikie concluded, Scotland had once been covered in ice!

U-shaped valleys are common on the landscape of Central New York, where I live. Early researchers realized that the U shape indicated the former presence of flowing ice, transforming V-shaped valleys, previously eroded by running water, into the rounder outline. An "ice river," more ponderous than flowing water, cuts steeper walls and smoother floors, forming the characteristic U.

Another sign of glacial presence is the *drumlin*, a hill of soil covering a loose mix of rocks and gravel, shaped like a spoon with convex side upward. These formations were produced by glaciers that moved around them, the hills aligned with the direction of glacial flow, their north slope fairly gradual while the south slope is steeper. I am writing this in Syracuse, New York, where my home is built on the south (steep) side of a drumlin that offers a lovely view of neighboring drumlins (Figure 4.5). Around Boston, the Revolutionary War sites of Bunker Hill and Breed's Hill are drumlins.

Eskers are landforms produced by glacial meltwater discharge, formed when stream tunnels under the ice become filled with gravel and rocks. After the glacier melts, the tunnel fill remains as a winding ridge.

Figure 4.5 Neighboring drumlins, from my study window, Syracuse, New York

There are also *kettle holes*, some quite large, formed by blocks of ice broken off from the retreating sheet. Meltwater carried sand from the glacier's edge, leaving outwash around these remnant ice blocks, so when they finally melted, they left a kettle-shaped depression in the ground. One of my favorite picnic destinations, Clark Reservation, is a large kettle hole, a 15-minute drive from my house. Better-known kettle holes are the freshwater ponds of Cape Cod.

With the former extent of ice sheets, and their directions of flow, best indicated by striations in the bedrock – also by locating the source of erratic rocks "out of place" where they now lay, maps could be drawn of former great ice sheets and their points of origin. These disproved Agassiz's original notion that a single ice sheet had spread out from the North Pole to cover most of the northern hemisphere. Instead, individual ice sheets expanded from different centers. In North America, for example, the Laurentide Ice Sheet spread from a center near today's Hudson Bay, flowing northward toward the shore of the Arctic Ocean, then, as now, apparently covered by relatively thin floating ice; it flowed south to roughly today's border between New York State and

Figure 4.6 Chamberlin's map of the Laurentide ice sheet over east Canada, separated from the Cordilleron ice sheet in western Canada. Note that Alaska is unglaciated, and there is an ice-free corridor between the Cordilleran and Laurentide ice sheets, which may have allowed the first Americans to migrate into North America from Asia.
Source: Geikie 1894: Plate XIV

Pennsylvania (Figure 4.6). Eventually Europe's Alpine glaciation was discerned to be localized to that mountainous landscape, unattached to the large Scandinavian ice sheet descending from the North.

The height of an ice sheet could be measured by the height of mountains that had been smoothed and striated to the top, and especially from higher mountains that had been marked to the height of the moving ice but left rough and uneven above it. This produced estimates that the northern continental ice sheets were a mile thick.

Almost immediately it became apparent to others, if not to Agassiz, that an implication of the glacial theory was that an enormous volume of water had been removed from the oceans to build up a mile-high ice sheet on land, so that sea level would have decreased by hundreds of feet, opening land bridges such as connected Asia and Alaska. Also, the enormous weight of the ice sheet would have depressed the underlying land, so when the ice melted that land would gradually rebound, increasing the apparent drop in sea level. This explained the presence in previously glaciated areas of marine fossils above present-day sea level.

There were indications in America and Europe that glaciation has occurred not once but multiple times. This was indicated by two or more layers of till deposits separated by layers with plant fragments, indicating that a warm climate had intervened between different glacial ages. Today we are living in a naturally warm interglacial, apart from our own contribution to the warming climate.

There were new insights about areas not themselves covered by ice but in neighboring regions. Large parts of Europe, Asia, and North America had been blanketed during the Ice Age with a layer of fine, homogeneous, yellowish, unstratified sediment called "loess," sometimes in deposits exceeding ten feet (3 m), and with no sign of water deposition. When melting occurred at the southern boundary of the ice sheet, great quantities of silt were deposited by outwash streams, and these were easily blown away, producing fertile soil for new grasslands, as in the American Midwest.

> By 1875 geologists had completed their initial survey of what the world of the last ice age was like. They had mapped its glaciers; measured its sea level; and determined which areas had been cold and wet, which cold and dry. They had also discovered that the ice age was not a unique event – that, in fact, there has been a succession of ice ages, each separated by warmer, interglacial ages similar to the present one. (Imbrie and Imbrie 1978: 57)

In 1894, Thomas Chamberlin (1843–1928) of the University of Chicago made a first attempt at mapping the extent of the last ice age in North America, published by Geikie in the 1894 edition of his book (Figure 4.6). Modern mapping of Northern Hemisphere ice sheets 18,000 years ago roughly verifies his attempt, though adding another expanse of ice centered on the islands of the Arctic Archipelago and connecting to the Greenland ice sheet (Figure 4.7). Today it is recognized that mountainous regions – the Alps, the Rocky Mountains, the Sierra Nevada, the Andes, the Himalayas – nucleated their own glaciers, separate from the great ice sheets. Thus, the Alpine glaciers that Agassiz avidly explored were not part of the continent's great northern icy expanse, as he thought,

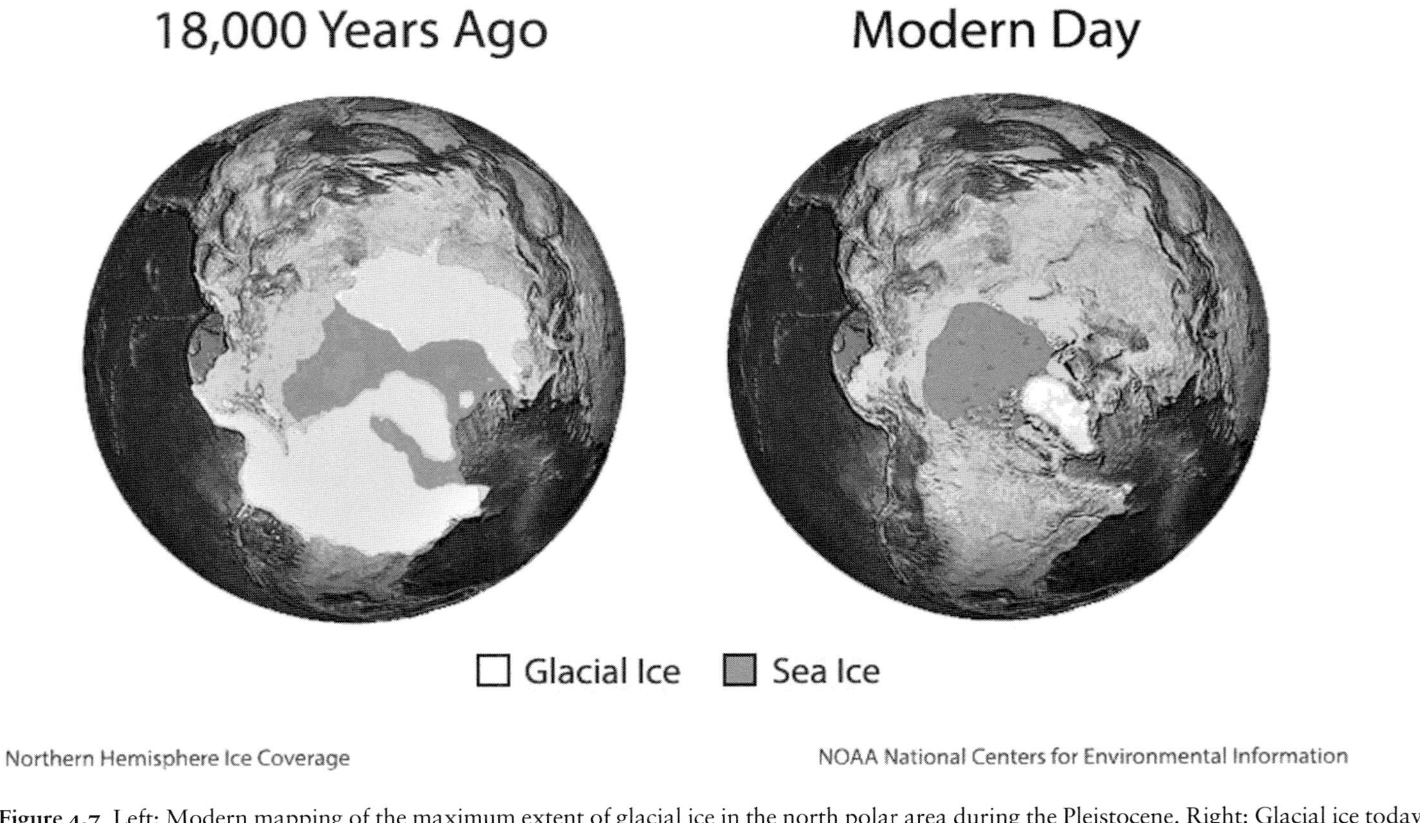

Figure 4.7 Left: Modern mapping of the maximum extent of glacial ice in the north polar area during the Pleistocene. Right: Glacial ice today. *Source*: NOAA

but were a separate glacial area, south of Eurasia's Scandinavian ice sheet and unconnected to it.[1]

If one's eyes are attuned, signs of past glaciation are abundantly clear. It is a rare visitor to New York City, the most popular tourist destination in America, who comes to see remnants of an Ice Age, but one could hardly do better. The Laurentide ice sheet, reaching a maximum depth of two miles, began receding 18,000 years ago, leaving a great terminal moraine across the country. In the east, it runs through Manhattan, Brooklyn, and Staten Island. The moraine's maximum height today is roughly 200 feet (60 m) and is visible in New York as an elevated band of hills, parks, golf courses, and cemeteries. Erratic boulders in Central Park are favorite climbing spots. The park's exposed bedrock is deeply striated. In Brooklyn, Prospect Park was built on the ridge as well as beautiful Greenwood Cemetery, site of the 1776 Battle of Brooklyn, the first war of the American Revolution. Fashionably gentrified Park Slope is named for its position on the moraine. Staten Island and Brooklyn were once connected by the moraine, but glacial floodwaters running down the Hudson River burst through 13,000 years ago, forming today's Narrows (Broad 2018). The Empire State Building and some other skyscrapers, had they then been present, would have risen above the snow, but most of Manhattan's structures would have been buried.

[1] See the YouTube video "The Geography of the Ice Age" for the extent of ice, ocean, and ice-free land during the Last Glacial Maximum, 20–26 thousand years ago.

5 WHY DOES CLIMATE CHANGE?

Orbits

In 1874, the Scottish geologist James Geikie (1839–1915) recapped geological evidence for an extensive ice epoch and then addressed the unresolved matter of its causes:

> The stony record everywhere assures us that from the earliest times of which geologists can take cognizance down to the present, our globe has experienced many changes of climate. The plants of which our coal seams are composed speak to us of lands covered with luxuriant growths of tree-ferns and aur-acarians, and the fossils in our limestones tell us of warm seas where corals luxuriated in the genial waters. Nor is it only in our own latitudes that scenes like these are conjured up by a study of the rocks … Similar results have been obtained by many of our most distinguished arctic voyagers, and from their observations it is now well ascertained that over all the regions within the Arctic Circle which have yet been visited, genial climates have prevailed at different times during past geologic ages – climates that not only nourished corals and southern molluses in the seas, but clothed the lands with a rich and luxuriant greenery. (94)
>
> … The old belief used to be that the climates of the globe, owing in great measure to the escape of the earth's internal heat into space, had gradually and regularly cooled down; so that the older formations were thought to represent ultra-tropical conditions, while the later deposits contained the records of less tropical and temperate climates … But the rapid accumulation

> of facts proved fatal to this as to many other theories ... The earth, no doubt, still radiates heat into space, but in considering the history of the past, so far back as it is revealed by geology, this cooling of the earth may safely be disregarded ... We must look elsewhere than to the secular cooling of our globe for the causes which have at several periods induced a mild and even genial temperature within the Arctic Circle – periods during which the whole northern hemisphere enjoyed a kind of perpetual summer. For we know now that such genial conditions had been preceded and were eventually succeeded by climates of more than arctic rigour, when our hemisphere, which had luxuriated in one long-continued summer, became the scene of great snow-fields and glaciers and floating ice. This alternation of genial climates with arctic conditions obviously cannot be accounted for by the cooling of the earth, due to the radiation of its central heat into space. (97)

Geikie reviewed attempts to account for these alternations in climate. There was speculation that the entire solar system passes through warmer or colder regions of space, thereby producing correspondingly warmer or colder climates on Earth. Some imagined a change in the position of Earth's axis of rotation. One "ingenious writer" proposed that the entire external crust of the globe slides around its fluid nucleus, so that lands that at one time bask under a tropical sun might at other times be near a pole, covered in ice.

The theory that Geikie thought had taken firmest hold was one by Sir Charles Lyell, who decades earlier had pointed out that if all lands were clustered near the equator, their surfaces would be tropical, whereas if these lands were clustered near the north and south poles, they would be frigid and covered with ice. Lyell's configurations are diagrammed in his 1834 edition of *Principles of Geology*. (See Figure 8.1.) He was not anticipating continental drift but thought such reconfigurations possible through the usual alternations of land and sea floor caused by erosion, subsidence, and gradual elevation. His proposal was that climate would change with different distributions of land.

Geikie, finding all these speculations implausible, argued that changes in climate – the long-term alterations between arctic conditions and "perpetual summer" – must be caused by slow changes in Earth's orbit around the sun. In this he would turn out to be partly correct. This view was not wholly original or uncontroversial. The astronomer Sir John Herschel had suggested that orbital variations might solve the geologists' puzzle of shifting climates, but other astronomers argued cogently that orbital changes could not be very important. Krüger (2013) provides more detail on early attempts at orbital explanations for Ice Ages. It is worth emphasizing that geologists and astronomers of that time had little contact with one another because they dealt with seemingly unconnected spheres of inquiry.

A critical point for Geikie and others was that "the" Ice Age had turned out to be a series of Ice Ages, separated by interglacial warmth, occurring over a very long time, so their cause or causes must have a repetitious character – it could not be a singular event. The periodicity of planetary obits fit that bill. (The term "Pleistocene" came into use to refer to that part of the Cenozoic Era in which there had been a sequence of glacial periods.)

To properly follow this debate, it is worth stepping back over two millennia for a look at the surprising sophistication of ancient astronomy. Readers who would rather not may skip ahead to the next section.

Ancient Astronomy

An intelligent but scientifically naïve human would have no way to decide if the alternation of day with night, or the fixed pattern of stars moving through the night sky, was caused by a spinning Earth revolving around the sun, or the sun and a celestial sphere revolving daily around a fixed Earth. Indeed, there was a simple physical argument favoring a stationary Earth for if it were rotating fast enough to account for the daily cycle, any unanchored object on its surface would fly off into space. Common sense dictated that the sun and celestial sphere were moving, not Earth.

The Bible barely weighs in on the matter, excepting one passage in Joshua (10: 1–15) where God makes the sun stop in the middle of the sky, lengthening the day so the Israelites' slaughter of enemies could be completed. This seems on its face to indicate a moving sun, but given the flexibility of interpretation, it could have referred to the *appearance* of the sun stopping because Earth stopped rotating (though if Earth's rotation suddenly stopped, Joshua would have flown eastward off the battlefield). Today, of course, no fundamentalists cite the Joshua story as evidence that the sun revolves around Earth, so that matter at least is settled to everyone's satisfaction.

Apart from whether Earth is moving or stationary, ancient astronomers knew it is a sphere. That the surface is curved is easily deduced from the sight of ships sailing toward the horizon, their hulls lost from view while their masts are still visible. That it is a globe was seen during lunar eclipses when Earth casts a circular shadow on the moon.

The astronomer Eratosthenes of Cyrene (modern Aswan, Egypt, ca. 276–195 BC) not only knew Earth to be spherical but is credited with a fairly accurate estimate of its circumference. He did this by comparing altitudes of the mid-day sun at Aswan and Alexandria, a known distance apart in a north-south direction. From these measurements, he calculated that the cities were separated by about 1/50 of Earth's circumference. Knowing the distance between the two cities, he calculated the distance around the total circumference.

There is a schoolyard myth that when Christopher Columbus was seeking backers for his westward route to the Orient, the savants of Europe thought Earth was flat and the project hopeless, which certainly is not true. Columbus's scholarly opponents, knowing the work of Eratosthenes, disputed his project primarily because of the distance that would have to be sailed around the world to reach Asia, which he put at 2,500 miles (4,023 km) and they put at 10,000 miles (16.093 km). They were closer to the truth (about 13,000 miles, or 20,091 km), but of course, no one suspected that two "new" continents blocked the route at about the distance where Columbus expected to find Asia. Columbus insisted to his death that he had reached the Orient, although in the end hardly anyone believed him.

The alteration of day and night, whether because Earth or the sun is spinning, has an axis of rotation that points toward the North Star, Polaris. The "equator" is defined as the plane to which the axis of rotation is perpendicular. This must not be confused with the plane of Earth's movement around the sun (or if one preferred, of the sun's orbit around Earth), known as the *ecliptic*. Greek and Hellenistic astronomers recognized that the day-night rotational axis was not perpendicular to the ecliptic but tilted away from that perpendicular at an angle of roughly 23½°.

This is why we have seasons. During winter in the Northern Hemisphere, Earth's axis of rotation is tilted about 23½° away from the sun, giving Northerners shorter days than in summer. In Northern summer, the rotational axis is tilted about 23½° toward the sun. It is the length of daylight, more sunshine in summer, less in winter, that gives us a seasonal change in temperature. The Southern Hemisphere is exactly opposite with summer in December, winter in June. (Equatorial latitudes do not have four seasons. At the poles there are six-month days and nights.) Many people assume summer is warmer because Earth is closer to the sun, which is incorrect. In our present era, Earth is actually closer to the sun in the North's wintertime than in summer.

Incredibly, the Hellenistic astronomer Hipparchus of Nicaea (ca. 190–120 BC) realized, perhaps by comparing his own observations to much earlier Babylonian records, that the axis of rotation is not permanently fixed in position – i.e., does not permanently point toward the North Star – but moves very slowly in a circle with a period of roughly 26,000 years, pointing at different times to different stars. To physicists, this is the same phenomenon as the rotation of the axis of a spinning top, called "precession." Apart from precession of Earth's rotational axis, its orbit slowly changes position. By the nineteenth century, astronomers recognized that combining these motions reduced the period of precession to about 22,000 years.

In the late Middle Ages, astronomers in Muslim, Hindu, and Christian cultures realized that the tilt of Earth's rotational axis is not constant at 23½°, but it too slowly changes. By the nineteenth century the rotational tilt was known to vary between 22 and 24.5° on a 41,000-year cycle.

One can hardly fail to be impressed by the precision of premodern astronomers (and astrologers), even if most still thought Earth was the center of the universe. The Catholic Church insisted on this doctrine well into the Renaissance. Its geocentrism was not importantly based on the Bible, already noted, but on the medieval Scholastic Movement led by Thomas Aquinas (1225–1274), who tried to reconcile Scripture with Aristotelian logic on the presumption that both paths to Truth must end up at the same place.

Aristotle accepted a stationary Earth at the center of a revolving sun and celestial spheres. As this system was perfected by Ptolemy (ca. 100–170 AD), it was fully capable of predicting the positions of the known planets among the stars, and lunar and solar eclipses.

All ancient and medieval models of the solar system assumed that movement of heavenly bodies was circular because circles are the most perfect shape. Orbits were circles, and in the sophisticated system of Ptolemy included supplementary circles (epicycles) moving along the main circles. Even Copernicus (1473–1543), credited with placing the sun at the center of the solar system, used epicycles to make his calculations, which were not more accurate and only slightly simpler than those of Ptolemy. It is uncertain whether or not Copernicus, a devout churchman, actually believed the sun is at the center of the solar system. His book, coming out as he lay on his deathbed, carries a preface stating that the heliocentric system was just a calculational device, but historians commonly believe that this was penned by a friend to avoid a conflict with Church authorities. Copernicus, by the way, retained the ancient belief that orbits were circular, or a combination of circles on circles.

There is no doubt that Johannes Kepler (1571–1630), another churchman, believed the sun was truly at the center of the solar system, not because of any empirical evidence but because he thought it a more aesthetic arrangement (Koestler 1964). More relevantly, it was Kepler who first showed that planetary orbits are not circles but ellipses, with the sun at one focus. (Kepler also retained ancient ideas, believing there was a physically real celestial sphere in which the stars were embedded; he calculated the musical notes sounded by this sphere rotating on its heavenly axis.)

The degree to which an ellipse departs from a circle is measured by its *eccentricity*. Enlightenment astronomers realized that over thousands of years, the shape (eccentricity) of Earth's orbit changes, from being nearly circular to somewhat elongated, then back toward circularity, with a period of about 100,000 years. This is the third variation in Earth's orbit to remember. So we have (1) precession of Earth's rotational axis, on a 22,000-year cycle; (2) change in axial tilt, on a 41,000-year cycle; and (3) change in the eccentricity of the ellipse, on a 100,000-year cycle. With these in mind, we can examine the development of an orbital theory of the ice ages.

Croll's Theory

Scotsman James Croll (1821–1890), son of a stonemason and lacking formal education, produced the first serious theory of climate change based on variations in Earth's orbit (Fleming 2006; Gribbin and Gribbin 2015). He started his working life as an apprentice wheelwright, then moved through jobs as a merchant, hotel manager, insurance agent, and janitor at the Andersonian University in Glasgow, where he had access to the library's science books and taught himself physics and astronomy. His correspondence with Charles Lyell about links between Ice Ages and variations in Earth's orbit led to a job with the Geological Survey of Scotland, under the direction of Archibald Geikie (elder brother of James Geikie, who opened this chapter). In following years, Croll produced several publications emphasizing orbital mechanics as a key to climate change, especially *Climate and Time* (1875), gaining considerable attention, perhaps through the interest of the Geikie brothers. In due time, Croll was elected a Fellow of the Royal Society, quite an ascent from his mean origins.

The nub of Croll's theory is that decreases in winter sunlight would favor snow accumulation, and this would occur through variations in Earth's orbit. He pointed out that the difference in heat received by the two hemispheres would change with the eccentricity of Earth's elliptical orbit, increasing when the eccentricity reached its highest value, i.e., when the orbit was most oblong. If at a period of maximum eccentricity, the Northern Hemisphere's winter happened when it was farthest from the sun, it should receive less heat in that season than if the orbit were nearly circular. He also took into account the precession of Earth's rotational axis. These two orbital effects, taken together, provided a mechanism for multiple glacial epochs and alternating cold and warm periods in each hemisphere. A consequence was that when one hemisphere was in an ice age, the other would be in a warmer interglacial. (It was not known at the time whether the Northern and Southern hemispheres experience cold periods simultaneously or in alteration.) Using orbital parameters available from French astronomers, Croll calculated variations in Earth's orbit for three million years back and one million years in the future, showing irregular but predictable changes in terrestrial climate for both past and future.

Croll understood that terrestrial feedback effects were also important, including albedo, whereby extensive snow and ice fields reflect back more sunlight than would darker surfaces. He also understood that the deflection of ocean currents could alter heat transfer from the equator to the poles.

By the end of the ninenteenth century, as Ice Ages in the Northern Hemisphere were better – though still very roughly – dated, Croll's calculations seemed to be incorrect and were ignored. Eventually it was determined that glaciation in the Southern Hemisphere, more limited than in the North because there is less land on which ice can accumulate (obviously excepting Antarctica), roughly coincides with cold periods in the North, thus contradicting Croll's assertion that the two hemispheres alternate warm and cool periods. Croll's lasting contribution was to emphasize the importance of orbital mechanics, though it would be nearly a century before an orbital theory and the dating of ice ages became concordant.

Milankovitch's Theory

Serbian Milutin Milankovitch (1879–1958) completed a PhD in 1904 at the Institute of Technology in Vienna, then worked five years designing concrete structures while yearning for more fundamental problems. In 1909, he took a position as professor of applied mathematics at the University of Belgrade, a provincial school from the perspective of Vienna, but it was a return to his homeland that freed him to select his own research. There he chose to develop a mathematical theory of climates on Earth, Mars, and Venus, comparing the sun's effects on planets with different orbits. Aware of Croll's work and its inadequacy, Milankovitch thought he could do better as a trained mathematician and with better data. It was a problem to which he could devote years of his academic career. Nearly 30 years, as it would turn out, with Earth his major focus, especially the variations in past climate that might account for the succession of glacial and interglacial periods.

In the opening phase of his research, Milankovitch found, as Croll had before him, that three orbital properties determine how solar radiation is distributed over the planetary surfaces: the eccentricity of the orbit, the tilt of the axis of rotation, and the precession of that rotational axis. Croll had had at his disposal only calculations of historical variation in eccentricity and precession. Milankovitch could use astronomical calculations that had since become available on variations in all three key properties (eccentricity, precession, and tilt) over the past million years. From these he could calculate how much solar radiation struck the surface of each planet during each season and at each latitude (Imbrie and Imbrie 1979: 100). Before computers, these were enormously time-consuming and error-prone calculations.

In 1920, Milankovitch published his results to date, demonstrating that orbital variations were sufficient to produce Ice Ages by changing the seasonal distribution of sunlight. This book, though largely unnoticed, did reach the attention of the German climatologist Wladimir Köppen, who invited Milankovitch to collaborate with him and his younger nephew, Alfred Wegener (already known for his theory of continental drift) on a book about climates of the geological past.

One fruit of their collaboration was to select the season that was critical for the growth of an ice sheet. It was previously thought by Croll and others that winter was key, when snow would accumulate. Köppen pointed out that variation in winter temperature of the Arctic, as long as it stayed below freezing, would have little effect on the amount of snowfall. It was the amount of heat during the *summer* that was the decisive factor. Glaciers melt in summertime, so a decrease in solar radiation then would slow melting, facilitating the long-term buildup of ice.

Milankovitch laboriously calculated how summer radiation at 55°, 60°, and 65° north latitude varied over the past 650,000 years, then mailed the graph to his German colleague. Köppen responded that low points in the summer intensity of solar radiation matched reasonably well with the history of Alpine glaciers, to the extent that this had been crudely estimated. The graph was included in the book published by Köppen and Wegener (1924).

Continuing his calculations for lower and higher latitudes, geologists now understood how two of the astronomical cycles influences solar radiation. As Croll had foreseen, a decrease in axial tilt causes a decrease in summer radiation; a decrease in the Earth-sun distance causes an increase in radiation. The influence of the tilt cycle, the regular 41,000-year oscillation of the inclination of Earth's axis, is large at the poles and becomes small toward the equator. In contrast, the influence of the precession cycle, a 22,000-year oscillation of the Earth-sun distance, is small at the poles and becomes large near the equator. The inadequacies of Croll's theory, according to Milankovitch, were that he had not given sufficient importance to the variability of axial tilt, and his concentration on winter rather than summer temperatures. An inadequacy in Milankovitch's formulation, as we shall see, was that he gave insufficient emphasis to the 100,000-year cycle of eccentricity.

Milankovitch went on to estimate how much ice sheets would respond to a given change in solar radiation (insolation), and how the latitude of the margin of the ice sheet moved during the past 650,000 years. Publishing his full results in 1938, he completed his long research program a year before the outbreak of the Second World War.

6 DATING ICE AGE CLIMATES

The orbital calculations of Croll and Milankovitch gave dates for glacial episodes. Croll put the end of the last Ice Age at 80,000 years ago, Milankovitch at less than 25,000 years ago. But while astronomers provided numerical ages, geologists and paleontologists could give only relative dating. The history of life, as shown by fossils, was importantly marked by the succession of the Paleozoic (old life), Mesozoic (middle life), and Cenozoic (new life) eras. These eras were subdivided into smaller units called "periods," then even smaller units called "epochs."

During the nineteenth century, identification of periods and epochs was in flux because of new discoveries, competing claims, and differences between European and American geology. Boundaries between subunits were based on the appearances or disappearances of certain fossils, but agreement about which fossils to use and from which exposures was yet to come, and no one knew how many years ago these epochs were.

The term "Pleistocene" (from the Greek for *newest*) had been coined by Charles Lyell in 1839 to label strata in Sicily that had at least 70 percent of their molluscan fossils still alive, thus distinguishing them from a slightly earlier (Pliocene) epoch. But with the recognition of repeated glaciations, the meaning of "Pleistocene" was wholly redefined as the epoch of successive advance and retreat of great ice sheets, or colloquially, the Ice Age. We are presently in the Holocene epoch, a relatively short span of time (so far) following the Pleistocene epoch (or Ice Age), both included in the Quaternary period of the Cenozoic era. This terminology of these intervals can be confusing, so it is worth a diagram of those names used in this book, along with recently assigned dates (Figure 6.1).

German geologists concluded there had been four major glaciations, which they named for river valleys, from earliest to most recent:

Epoch	Glacial/interglacial periods	Includes
Holocene, 11.7 kya to present	Present interglacial	Agriculture, from ca. 10 kya Little Ice Age 1300 to 1850 AD
Pleistocene (Ice Age), 2.6 million to 11.7 kya	Last glacial period, 110.0 to 11.7 kya Last prior interglacial cycle, 130 to 110 kya Prior glacial and interglacial cycles, 2.6 million to 130 kya	Younger Dryas 12.9 to 11.7 kya Last glacial maximum 20 to 26 kya Near Time 50 kya to present

Figure 6.1 Named intervals used in this book(kya = thousand years ago)

Günz, *Mindel*, *Riss*, and *Würm*. Americans also thought there were four major periods of ice, which they named for the states in which these were most easily studied, from earliest to most recent: the Nebraskan, Kansan, Illinoian, and Wisconsin glaciations. A tempting inference was that these were the same intercontinental events under different names, but frustratingly, there was no way to correlate the American and European glaciations. Today the correlation is recognized as inexact, and more importantly, there were far more than four alterations of glacial and interglacial episodes – more like two dozen. Furthermore, there are variations in temperature *within* glacial periods. Certainly the Ice Age was not a persistently stable period of cold.

Nineteenth-century geologists made diverse and sometimes ingenious attempts to assign real dates to ancient events. None were trustworthy. Often the amount of time that had passed between conformities in a stratigraphic pile was guessed from the depth of sediment between them, whether very deep or very shallow, but obviously this made no allowance for changes in deposition or erosion rates.

Niagara Falls was used to estimate the number of years since either Noah's Flood or the end of the last ice sheet, depending on the beliefs of the investigator. The Niagara River, part of the border between New York State and Ontario, Canada, is 36 miles (58 km) long, flowing north from Lake Erie to Lake Ontario. Niagara Falls is about midway down the river. For the river's first 15 miles (24 km) it slopes gradually, running fairly calmly until its tumultuous drop over the famous precipice into a deep gorge. The highly erosive power of the falls continually pounds away at the lip, leaving huge piles of rock debris at its base. Rock falls are frequently witnessed by visitors and residents.

It is easy to imagine that the falls were once much closer to Lake Ontario, with erosion causing it to move gradually upstream and carve out the gorge below. By estimating the rate of recession, one ought to be able to calculate in years the present age of Niagara Falls, at least on the assumption that the rate of recession is constant. Robert Blackwell, Jr., did just that in 1829, estimating from local reports that the cataract moved back about one yard (ca. I m) annually, so Niagara Falls was roughly ten thousand years old, comfortably close to the time limit allowed by Genesis.

Charles Lyell made a prolonged visit to study Niagara Falls in 1841–42 and estimated the rate of recession was more like one foot a year, thus tripling the age calculated by Blackwell to 30,000 years. Lyell had long opposed any literal interpretation of Genesis and by this time was a convert to the ice age theory. He recognized the Niagara region as having been glaciated, and that the Great Lakes would be at least partly a consequence of the great melting. Thus, he put forward his estimate of 30,000 years as a reasonable date for recession of the last ice sheet (Imbrie and Imbrie 1979; Breton 1992). Earlier Lyell had written that the rate of recession was unlikely to be constant, which should have undermined the whole exercise. Incidentally, modern dating of Niagara Falls is closer to Blackwell's than Lyell's age.

Relative Dating

Paleontologists did much better at piecing together a chronology with relative dating. As the association of certain fossils with particular stratigraphic horizons became better catalogued, these were called "index fossils," and geologists came increasingly to depend on them to mark the relative age of the stratum in which they were found. Index fossils were essential tools for paleontologists until the invention of numerical dating methods after World War II.

By 1909, there had been sufficient identification of index fossils for Professor Amadeus Grabau (1870–1946) of Columbia University and his former graduate student, Hervey Shimer (1872–1965), then-assistant professor at MIT, to publish their monumental *North American Index Fossils*, on invertebrates. Aimed particularly at the student of stratigraphic paleontology, its two volumes run nearly 1,800 pages, describing thousands of species for field or laboratory identification, each one indicative of a certain horizon, operationally a geological *period*, or subdivision of an *era*. For example, the Mesozoic Era (popularly the "Age of Dinosaurs") is divided into three successive periods, from the Triassic, through the Jurassic, to the Cretaceous.

The authors introduce Volume I with an overview:

> In general, it may be said that the more precise the required identification of a horizon the more limited must be the range of the fossil or fossils which are relied upon to indicate that age. Thus while trilobites as a class may be relied upon as indicators of Paleozoic age, being unknown above this, a certain group of trilobites alone will serve to indicate Cambric age, while a genus

> (*Olenellus*, or *Paradoxides*) serves to indicate the lower or middle Cambric respectively. Furthermore, a certain group of species of *Paradoxides*, as for example the species of the *P. eteminicus* type, serve to indicate a certain horizon in the Middle Cambric. (Grabau and Shimer 1909: 1)

Their catalog became an essential reference for North American geologists. Often revised and reprinted, its circulation outlasted the original authors.

Long-term changes in climate could be inferred from changes in the fossils from one stratum to another. Some organisms were typical of tropical environments, others from cold regions. Pits were dug or vertical cores drilled in river and lake deposits, or in loess, to compare species of pollen, or other small organisms, at different depths, knowing that some flourish in cold, others in warmth. Though still unable to assign numerical dates to such changes, the recognition that changes occurred, in what order, and over what areas, was impressive and sometimes surprisingly detailed.

By the early twentieth century, for example, examination of changing pollens beneath Scandinavian bog and lake sites had revealed that after the region's last great ice sheets had melted, there was a seemingly sudden return to ice-age conditions. This return, today dated as beginning 12,900 years ago and lasting around 1,200 years, was called the "Younger Dryas." Dryas refers to an index species, the alpine-tundra wildflower *Dryas octopetala*, which spreads in cold weather (Alley 2014). The period was called "Younger" to distinguish it from two preceding and seemingly less extreme returns to cold, the Older Dryas and the Oldest Dryas.

As land-based geologists were refining their dating methods, similar efforts progressed in the more difficult environment of the sea. In 1872, when the oceans were barely known below the surface, the British steamship HMS *Challenger* began a 3½-year voyage of scientific discovery, sailing around the world to collect plants, animals, and water samples, making soundings, and dredging the seafloor. Much of the bottom was covered by sediments, but these were not all the same. The scientists aboard *Challenger* found oozy sediments on the floors of temperate or tropical seas, containing limy remains of a type of plankton known as foraminifera (or forams). They found another kind of ooze in the colder waters of the Arctic and Antarctic, having different plankton known as radiolaria and diatoms (Figure 6.2).

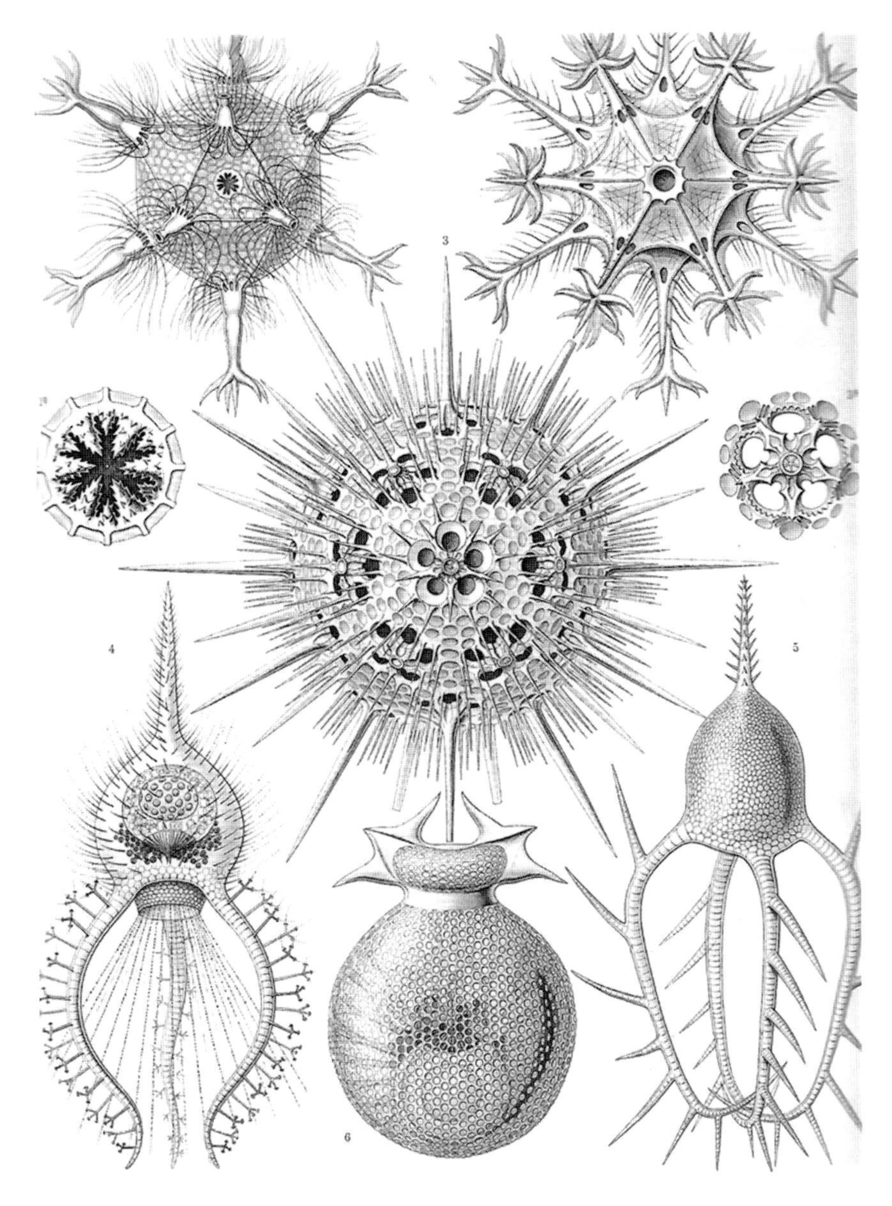

Figure 6.2 German evolutionist Ernst Haeckel's 1904 rendering of Phaeodaria plankton, traditionally considered radiolarians but not closely related in molecular trees. Haeckel was a marvelous artist, influential in the Art Nouveaux style.
Source: Haeckel (1904)

Since some kinds of plankton live in warm seas, while others live in cold water, there appeared a possibility of measuring changes in ancient climate from layers of sediment on the sea bottom. Like pollen analysis on land, tracing changes in the distribution of planktonic types, as they died and settled down to the bottom of the sea, would reflect rising or falling water temperature over time.

It was not easy to deeply penetrate bottom sediment with a pipe dangling by a cable from a ship on the surface, even using dynamite to drive the pipe into the sea bottom. A workable core collector capable of extracting one meter of sediment was eventually developed for the German ship *Meteor*, which in 1925–1927 collected samples from the floor of the equatorial Atlantic Ocean. In these, paleontologist Wolfgang Stott distinguished three layers, with the upper and lowest containing warm water foraminifera species, while the middle layer contained cold water species. Apparently the meter-long cores contained sediments that had settled before, during, and after the most recent cold period. Still no numerical dates could be set, but clearly there was a need for longer cores.

Numerical Dating

Henri Becquerel, in 1896, found that a photographic plate, wrapped in paper, was nonetheless exposed when placed near uranium, as if by some unknown emanation. Fairly quickly it was realized that uranium and certain other elements throw off previously unsuspected emissions, causing the plates to expose. This was called "radioactivity," which would become the key to numerical dating.

In the early twentieth century, the atom was pictured as a miniature solar system, its center a nucleus combining heavy protons and neutrons, this orbited by far lighter electrons. Protons and electrons are equally but oppositely charged, so the number of orbiting electrons is (usually) balanced by the number of protons in the nucleus, making the atom as a whole electrically neutral. It is the configuration of orbiting electrons that determines the chemical properties of an element, i.e., how its atoms will combine with other kinds of atoms. (Most elements were discovered because of their unique chemical properties.)

Each element is identified by its number of protons, its *atomic number*. The atomic number of uranium is 92 because it has 92 protons in its nucleus.

The neutron was pictured as a proton and electron welded together, thus electrically neutral. Since an electron weighs little compared to a proton, a neutron weighs barely more than a proton. The *atomic weight* of an atom is counted as the sum of its protons and neutrons in the nucleus.

A nucleus can gain or lose neutrons, i.e., change its atomic weight, without affecting the configuration of orbiting electrons, thus without affecting the chemical properties of the atom. However, gaining or losing neutrons affects the properties of the nucleus itself. For example, 99 percent of naturally occurring uranium atoms have 146 neutrons, giving an atomic weight of 238 (denoted uranium-238, or ^{238}U). Less than one percent of uranium has only 143 neutrons (^{235}U), making the nucleus far more unstable and the stuff of atomic bombs. The different versions of uranium are called "isotopes," meaning they differ in atomic weight but not in atomic number. Both are uranium, atomic number 92.

Sometimes a radioactive nucleus emits one or two protons, changing the atomic number (as well as the atomic weight) and thereby transmuting one element into another.

If one could watch a single uranium atom, there is no telling when its nucleus would decay, sending out an emission. The situation is entirely different when monitoring a large number of atoms. The rate of decay is constant and characteristic of the particular isotope. It is measured by its half-life, the time required for a quantity to reduce to half of its initial amount. The half-life of uranium-238 is 4.47 billion years, that of uranium-235 is far shorter: 704 million years. Both are useful in dating the age of Earth.

Radiocarbon dating, developed shortly after World War II by Willard Libby (1908–1980) at the University of Chicago, for the first time allowed numerical ages to be assigned to carbon-containing organisms. Libby had discovered that a radioactive form of carbon, the isotope carbon-14, is produced in small amounts in the atmosphere by cosmic rays. Since carbon-14 weighs slightly more than normal carbon-12, they can be separated by a technique called "mass spectroscopy."

Plants continually take up carbon-14 from the air while they are alive and photosynthesizing, and that carbon is passed up the food chain to animals, but on death this exchange stops. Since carbon-14 (but not carbon-12) is radioactive, whatever carbon-14 is in the organism at death will decay, eventually changing into inert atoms of nitrogen at a known rate, its half-life 5,730 years. To calculate the number of years since the death of the organism, one needs only measure the proportion of its total carbon that is radioactive (i.e., that is still carbon-14), and compare it to the proportion of carbon in the atmosphere that is carbon-14. Libby tested his method on wood samples from old trees, which can be dated independently by counting tree rings. These did not correspond exactly, but clever application of such methods allowed the calculation of calibration curves so that a radiocarbon date could be transformed to an actual number of years before the "present" (set at 1950). For example, he counted the number of rings in old trees, comparing their ages with his radiocarbon dates for the rings. With suitable calibration, radiocarbon dating works very well for samples less than about 50,000 years old. (After 50,000 years, too little ^{14}C remains for analysis.) This limited range did not carry measurement far back into the Ice Age, but still it was an enormous advance and there was a rush to apply it, sometimes recklessly for samples beyond the 50,000-year range. A new journal, *Radiocarbon*, was established to disseminate results.

Dating a large sample of peat, wood, shells, and bones from Wisconsin showed, as suspected, that there had actually been at least two glaciations there. The older samples were beyond the range of radiocarbon dating, but the younger pieces showed that Wisconsin's last great ice sheet reached its maximum extent roughly 18,000 years ago, then rapidly disappeared by 10,000 years ago. These dates were far too young for Croll's estimate that the last ice age ended 80,000 years ago, but they were fairly close to Milankovitch's date of $<$25,000 years ago (Imbrie and Imbrie 1979).

Now there was a profusion of published dates, not always consistent. In 1974 the International Commission on Stratigraphy (ICS) was established to standardize estimates based on the best available data; these remain in flux. In 2009 the ICS moved the start of the Pleistocene from 1.8 million years ago to 2.6 million years ago.

Development of similar techniques, based on uranium and other radioactive isotopes, or various trapped-electron methods (thermoluminescence, electron-spin resonance, or optically stimulated luminescence) expanded the types of material that could be dated (volcanic rocks, deep-sea mud) well past the half-billion years since the Cambrian Explosion. These were applied widely, including to cores extracted from the ocean floor, and from the ice sheets of Greenland and Antarctica.

Cesare Emiliani (1922–1995) at the University of Chicago developed a method for ocean core analysis based on isotopic composition of oxygen atoms in fossil forams. Seawater contains two isotopes of oxygen, O-16 and O-18, chemically identical, but O-18 is heavier and therefore less likely to evaporate and later fall as rain or snow. At the time, isotope chemists were finding that ice sheets contained high concentrations of O-16. Emiliani and others theorized that in cold periods when glaciation increased, the lighter atoms of O-16 were preferentially extracted from the sea and stored in land ice, increasing the ratio of O-18 to O-16 in the ocean. When glaciers melted, their isotopically lighter oxygen returned to sea, restoring the isotope ratio to its prior composition. Forams absorb oxygen from the sea, so their fossil shells preserve the ratio of O-18 to O-16 in the sea while they were alive. By comparing the isotopic ratio of forams living at different times, one could estimate changes in the temperature as well as the amount of water locked up in ice sheets on the land. The technique would become validated and is now routine:

> Take a drill ship to sea, and pull up a core of the sediment. Pay some poor student or technician to sort through the mud and pull out the shells of your favorite "bug" type. Use some of the great range of dating techniques to assign ages to the shells. Run the shells through your local mass spectrometer to measure the ratio of heavy to light oxygen. The result is a record of the size of ice sheets on Earth. (Alley 2014: 93)

Inferred trends were not always consistent, in part because the dating of core slices was uncertain, and partly the difficulty of matching records from different places. Firm anchor points were needed to resolve such differences, and these turned out to be Earth's shifting compass points.

Nailing Down the Magnetic Poles

Compasses orient to Earth's magnetic field, pointing toward the *magnetic north pole*, which is not the same as the geographical north-south axis around which Earth rotates. The current location of the magnetic north pole is in Canada's Baffin Island, so a compass in northern Norway would actually point west, toward Baffin. Expeditions to the magnetic north pole, led by James Clark Ross in

1831 and Roald Amundsen in 1903, each found it in a different position. Such movement was long recognized.

At nearly the same time as Amundsen's trip, a less adventurous geophysicist named Bernard Brunhes (1867–1910), working in a French brickyard, examined new bricks coming out of the kiln and found that iron-rich particles in the hot clay had oriented in a magnetic north-south direction. He was testing the idea that iron particles in cooling lava might do the same thing. If so, erupting lava would preserve the direction of magnetic north at the time it cooled. In later comparisons of the directions of magnetization in several ancient lava fields, Brunhes was surprised to find that the direction of the field had actually reversed!

Two decades later, a Japanese geophysicist, Motonori Matuyama (1884–1958), also working on ancient lavas, found multiple reversals in the distant past. During the early Pleistocene, the direction of Earth's field was reversed from its present direction. The idea that compasses of the past would point in the opposite direction seemed bizarre and was generally ignored. But in the 1950s and 1960s, better funded geologists confirmed such reversals. By then, the more recent reversals could be dated using one of the offshoots of radiocarbon dating, the potassium-argon method, especially well suited for volcanic rocks. The ages of two of these reversals, 700,000 years ago and 1.8 million years ago, provided worldwide fixed points from which a global chronology of the Pleistocene could be constructed.

During the postwar decades, a considerable "library" of ocean cores was collected by oceanographic research ships. Several showed magnetic reversals. Their pattern allowed cores from widely separated regions to be correlated.

At the same time as these technically advanced techniques in dating were being developed, Czech geologists made remarkable progress with the old-fashioned method of digging in a quarry. The site, near the city of Brno, was unusually favorable, containing undisturbed deposits of windblown silt (loess) alternating with layers of soil. This region had not been covered by ice during the Pleistocene but still was greatly affected. When the ice sheet farther north was large, Central Europe was a cold desert, dry and treeless, its winds depositing layers of loess picked up from glacial margins and other exposed ground. When the glaciers receded, the climate of Czechoslovakia was warmer and wetter, forested and with good soil. As the ice sheets alternatively expanded and contracted, the boundary between dry prairie and forest moved back and forth in a cycle manifest in the alternating layers of loess and soil in the quarry walls. There have been at least ten repetitions of the soil-loess cycle.

In 1968, George Kukla (1930–2014) and his colleagues detected five magnetic reversals among their layers of soil and loess. From this, they estimated that a cycle took about 100,000 years, a periodicity not predicted by Milankovitch. Furthermore, it seemed the cooling phase took much longer than the warming phase. At nearly the same time, Jan van Donk and Wallace Broecker at Lamont, using isotope measurements of forams in a Caribbean core, found similar results, concluding that the major pulse of climate was a 100,000-year cycle, and that the cycle had a saw-toothed shape, with glacial expansion taking most of it and then a relatively abrupt deglaciation (Imbrie and Imbrie 1979).

Kukla soon joined the Lamont-Doherty Laboratory of Columbia University, an emerging powerhouse in research on ancient climates. His obituary from Lamont, dated June 6, 2014, describes his persistent belief that Earth is rapidly moving toward another Ice Age. He became a public spokesman for that position, featured in cover stories in *Time* and *Newsweek*. Holding adamantly to that position, even as evidence of global warming mounted, Kukla became popular among conservative political groups like the Heartland Institute, which opposes the notion that human activity is warming the climate.

These and other lines of work made clear by the early 1970s the importance of the 100,000-year cycle. Recall that Milankovitch's theory predicted cycles based on the 41,000-year cycle of axial tilt, and the 22,000-year cycle of precession. He knew that the eccentricity of Earth's orbit has a period of 100,000 years but minimized its importance. Kukla and Kenneth Mesolella, reverting to Croll's emphasis on eccentricity, modified the Milankovitch theory to incorporate the longer cycle. But this post hoc adjustment was clearly fudging to fit those data and did not initially elicit much support for the orbital explanation.

In 1976, a report in the prestigious journal *Science* finally provided solid footing for the Croll-Milankovitch notion that variation in Earth's orbit was the primary driver of waxing and waning ice sheets during the Pleistocene. Incorporating all three orbital parameters – axial tilt, precession, and importantly the 100,000-year period of eccentricity, the authors explained statistically the large variations in climate as measured in 450,000 years of ocean-floor sediments from the Indian

Ocean (Hays, Imbrie, and Shackleton 1976; Imbrie and Imbrie 1979). They also modeled future climate, based on their observed orbital-climate relationships but ignoring anthropogenic effects, predicting that the long-term trend over the next seven thousand years was toward extensive Northern Hemisphere glaciation – another Ice Age.

Ice Cores

More than 99 percent of Earth's ice lies in sheets covering the lands of Greenland and Antarctica. Ice coring during polar summers began in Greenland in 1950s, and in Antarctica over the next decade, usually collecting three-foot (ca. 1 m) cores, ca. 5 inches in diameter, which, when patched together run a mile or two down into the ice, recording tens of thousands of years of accumulated snowfall.

In Greenland, the continual sunshine of three-month summers causes surface snow to look differently than snow accumulating during the sunless winter. This leaves a seasonal signal that is visible in the core and may be dated by simply counting layers, like one would count tree rings. A year of accumulated snow, perhaps initially a meter deep, becomes compressed as the snow of later years accumulates above it, weighting it down and turning it to ice. Layers that fell thousands of years ago, far down the core, are thin but still visible for counting, if not with the accuracy of younger layers. Counting can be instrumented, based on physical changes that occur with the changing seasons, for example, there is a yearly cycle of electrical conductivity of the ice (Conkling et al. 2011; Alley 2014; Gertner 2019).

Unusual events like volcanic eruptions leave recognizable depositions in the core. If the date of a large eruption is known from historic sources, that provides a validity check. Dust and sea salt in the ice were transported to Greenland by wind, and change in their concentration tell about changes in the winds. Bubbles of old air formed in the ice, showing the composition of the atmosphere at that time.

Records of multiple cores must be combined to infer the climate history of a broad region. These cores show that ancient climate had larger, faster, more widespread changes than any experienced by agrarian or industrial humans. Greenland cores going back 110,000 years show the last glacial period beginning with a 90,000-year decline from a warm time like the present into the cold, dry, windy conditions of an Ice Age, its maximum depth dated about 20,000 years ago, followed by a rapid return to warming, causing retreat of the great ice sheets.

There are deeper cores from Antarctica where the Russian Vostok project went more than three kilometers into the ice, providing usable information for the past 400,000 years. Figure 6.3 from Vostok data shows four climate cycles, each about 100,000 years, with warming occurring faster than cooling, giving a saw-tooth appearance to the temperature graph (Petit, et al. 1997; Petit et al. 1999). The European Project for Ice Coring in Antarctica (EPICA) has reached nearly to bedrock in the continent's interior, continuing the record to 800,000 years (Jouzel et al., 2007; Jouzel 2013).

Positive feedbacks are the reason that deglaciation occurs faster than glaciation. As changes in Earth's orbit increase summer insolation,

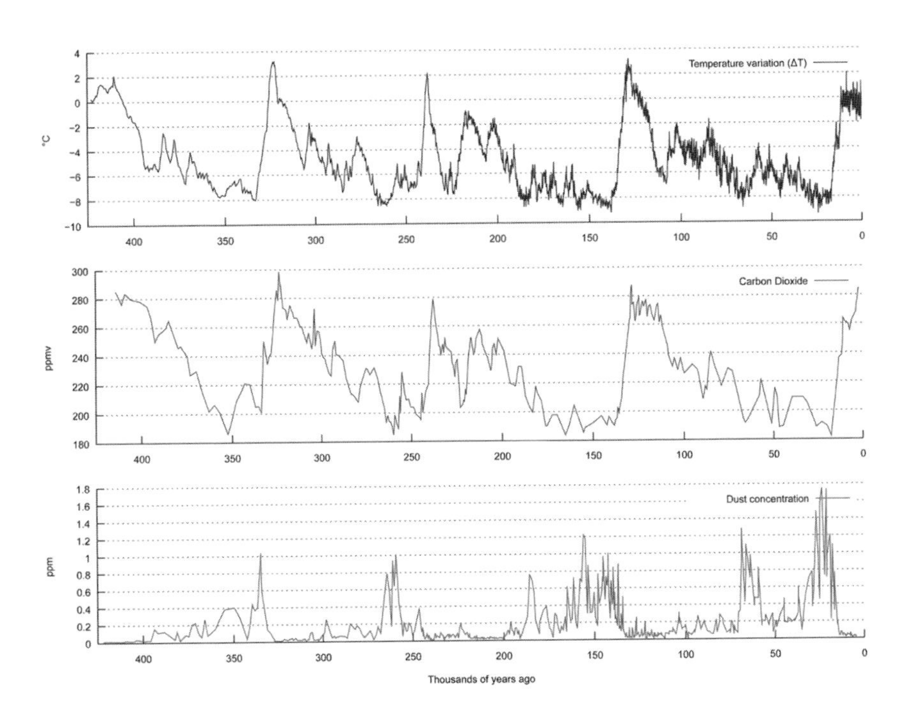

Figure 6.3 Graphs of reconstructed temperature (top), CO_2 (middle), and dust (bottom) from Vostok ice core for the past 420,000 years, showing 100,000-year cycles. Croll-Milankovitch cycles drove the ice ages; changes in CO_2 slightly lag temperature changes.
Source: Modification of Vostok-ice-core-petit.png, NOAA, from Wiki Commons, under GNU Free Documentation License

the warming sea releases CO_2 into the atmosphere, strengthening the greenhouse warming effect. Rising sea levels from melting glaciers undermine the margins of ice sheets, replacing surface areas of white snow or ice with dark areas of sea, a reduction in albedo that also amplifies warming (Woodward 2014).

Ice cores show irregular, shorter-term changes with abrupt warmings and coolings, unrelated to orbital variations. The best known of these, the Younger Dryas event, began 13,000 years ago, as the great northern ice sheets were regressing. This cold snap, nearly returning Earth to ice-age conditions, ended 11,500 years ago with Greenland warming about 15° F in a decade or less (Alley 2014). Then the prior warming trend resumed, bringing our recent 11,000 years of hospitable climate stability.

Greenland ice cores also reveal a remarkable series of more abrupt warming and cooling cycles, called "Dansgaard-Oeschger" (D-O) events after their discoverers. There were 25 such events between 65,000 and 10,000 years ago, when temperatures on the Greenland ice sheet warmed between 5° and 10°C within decades, then fell more gradually. Climatologist Hartmut Heinrich found six distinctive spikes in North Atlantic sediment records, not obviously related to D-O events, but again showing spikes of significant warming, then cooling. The causes of D-O and Heinrich events are unknown, as are their significance for global climate, but they further belie any notion that the Ice Age was a period of stable or slowly changing temperature and aridity. To the contrary, they show that significant changes in climate have occurred within one human lifetime and may again, even without anthropogenic tampering.

One would like ice cores that go deep enough, with annual layers that remain discernable enough, to trace climate through the entire Pleistocene and even earlier. Attempting to overcome this limitation, geologists combine many cores of ocean sediments, formed during different periods of the Pleistocene, into a single stack. Each core's place in the stack is determined by matching its pattern of short variations to patterns in potential neighboring cores. One of these stacks, based on 57 globally distributed ocean cores, represents the span of 5.3 million years, over ten times that from Vostok ice (Lisiecki and Raymo 2005). As before, change in the concentration of the oxygen isotope O-18 measures increases and decreases in the volume of glacial ice.

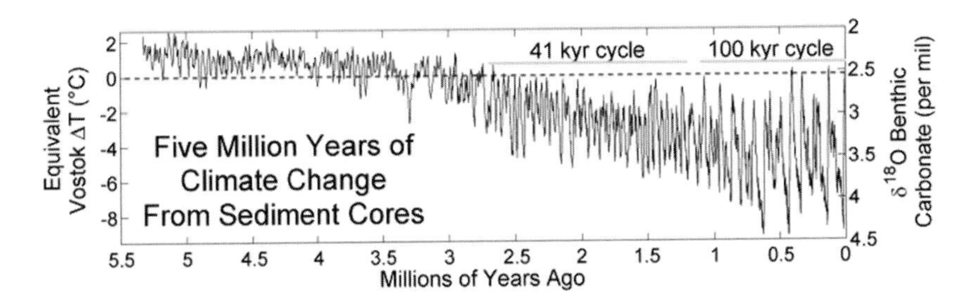

Figure 6.4 Five million years of climate change, including the entire Pleistocene epoch, shown by a O-18 proxy for the total global mass of glacial ice sheets. About one million years ago, the dominant glacial periodicity changed from 41,000 years per cycle to 100,000 years per cycle. The right axis shows change in O-18 concentration in forams, the left axis shows temperature change as an equivalence to Vostok measurements.
Source: Wikipedia, Dragons flight (Robert A. Rohde)

In these long-duration stacks, one can see a major alteration in the timing of glacial cycles from the early to the late Pleistocene (Figure 6.4). The changeover point, termed the Mid-Pleistocene Transition (MPT), is dated approximately one million years ago. Since then, the dominant glacial cycle was 100,000 years, corresponding to the orbital period for eccentricity. Earlier, before the MPT, the dominant cycle was 41,000 years, corresponding to the orbital period for obliquity). The reason for this transition is unexplained.

7 WHY DOES CLIMATE CHANGE?

Carbon Dioxide

Milankovitch cycles can change sunshine at a particular place and season by more than 10 percent, driving changes between glacial and interglacial periods. Reduced sunshine in the Northern Hemisphere summer allows the growth of northern ice sheets, while increased sunshine in the Northern summer causes its ice sheets to recede. Orbital change affects the *distribution* of sunshine but not the *total amount* reaching our planet from one year to the next. Yet when the Northern Hemisphere glaciates, the whole planet cools, with ice growing in the Southern Hemisphere too. When the North warms, so does the South. Orbital cycles are insufficient to explain this worldwide synchrony.

Changing intensity of sunlight was initially a candidate for the main driver of global-level climate cycles, but today that seems implausible. Modern solar physics explains the sun's heat arising from the fusion of its hydrogen atoms into helium, under the sun's enormous gravitational pull, a process that slowly makes the sun hotter and would not reverse to produce cold periods. Sunlight does vary a bit, notably over the 11-year sunspot cycle, but this is inconsequential for long-term climate (Haigh 2007).

Looking for earthly causes of changes in global climate, one possibility already recognized in the nineteenth-century was changing composition of the atmosphere. Among the first to pursue this idea was Irish physicist John Tyndall (1820–1893), son of a policeman, who first worked lucratively as a draftsman and surveyor for the burgeoning railroads. In the 1840s, he moved to Germany for its excellence in science education, there attaining a doctorate in science from the University of Marburg and training as an experimentalist. Returning to Britain in 1851, he attracted the support of the great Scottish experimental physicist Michael Faraday and was elected the next year to the Royal Society.

Perhaps influenced by German Romanticism, Tyndall became a pioneering mountain climber, visiting the Alps repeatedly, scaling its peaks and pondering its glaciers. He might be represented, like Agassiz, by Caspar David Friedrich's 1818 painting "Wanderer above the Sea of Fog" (see Figure 2.1). Puzzling like others of his time over the causes of ice ages, he thought they could be due to the changing gas composition of the atmosphere, rather than to orbital mechanics or the long-term waxing and waning of sunshine.

Greenhouse ("Hot House") Gases

In 1859, as a professor of physics at the Royal Institution of Great Britain, Tyndall began experiments on how infrared radiation, then called "radiant heat," passes through various gases in the atmosphere. He demonstrated that simple gases – oxygen, nitrogen, and hydrogen, which comprise most of the atmosphere – are transparent to infrared rays, allowing them to pass without hindrance. However, more complex molecules, specifically carbon dioxide and especially water vapor, absorb infrared radiation, in effect blocking its passage.

Think of a sunbeam of visible light entering the atmosphere, unimpeded by any molecule encountered during its inbound journey. When it hits Earth, the sunbeam gives up some of its energy as heat, then reflects toward outer space, no longer as visible light but as thermal radiation with less energy. On its outward journey, if it collides with a molecule of water or carbon dioxide, the infrared beam cannot pass. Metaphorically speaking, water vapor and carbon dioxide are one-way gates, allowing inbound visible sunlight to pass through, but blocking outbound infrared radiation from leaving. By retaining this heat – i.e., not allowing the infrared beam to escape into space – water vapor and CO_2 cause heating of the atmosphere. If it were not for the heating effect of these naturally occurring "hot house" gases, as they became called, Earth's atmosphere would be too cold to support life.*

* An American woman, Mrs. Eunice Foote (1819-1888), could be recognized among the founders of climate science, but she is not. Foote's 1856 paper, read (by a man) to a

Tyndall emphasized water vapor, not CO_2, as most important in affecting atmospheric temperature because of its greater power to absorb infrared radiation (Tyndall 1864; Fleming 1998). Carbon dioxide seemed less relevant in comparison.

The Swedish physicist and chemist Svante Arrhenius (1859–1927) proposed in 1895 that fluctuations of atmospheric carbon dioxide drove global temperature change sufficiently to cause glaciations and interglacial periods. He estimated that a reduction of atmospheric CO_2 to one half its current level would reduce temperature about 2°C, and a reduction to one fourth would cool by 4 degrees, sufficient to bring on glaciation. Doubling the level of CO_2 would produce warming of 2°. (Fortuitously, this roughly agrees with modern climate models for a doubling of atmospheric CO_2 over preindustrial levels.) Arrhenius predicted that the enormous increase in the combustion of coal would increase warming to a perceptible degree, thereby allowing humans to enjoy relief from what he thought would be an otherwise cooling climate in the future (Arrhenius 1908).

Around 1900, Arrhenius became involved in setting up the Nobel Prizes, using his position to award prizes for his friends and deny them to his enemies (Coffey 2008). In 1903, Arrhenius himself was awarded a Nobel for work unrelated to climate: his theory of electrolytic dissociation, i.e., that solid crystalline salts disassociate into paired charged particles when dissolved.

Ideas about atmospheric effects on climate were taken up in America by the distinguished geologist Thomas Chrowder Chamberlin (1843–1928), the same man who compiled the first map of the Laurentide ice sheet shown in Figure 4.6. He had grown up in rural Wisconsin, the son of a Methodist minister, imbued with a religiosity that predisposed him to see geological changes as driven by catastrophic deluges, consistent with biblical narrative. Eventually he rejected this view as he pursued his education in geology. Joining the Wisconsin Geological Survey, he was assigned to a portion of the state rich in glacial deposits and opportunity for plenty of fieldwork, during which

convention of the American Association for the Advancement of Science, described temperature changes in gases sealed in glass cylinders and exposed to the sun. Foote found that sunlight especially raised temperatures in cylinders containing carbon dioxide or water vapor. She conjectured that increases of these gases in the atmosphere could change the climate. Her work was ignored until recently brought to light (Jackson 2019). Technically speaking, Foote had not found what we regard today as greenhouse warming because she could not differentiate *incoming visible* sunlight from infrared radiation. As now understood, greenhouse warming comes from blocking *outgoing infrared radiation.* Still, Foote was on the right track, three years before Tyndall began his experiments.

he found evidence for multiple episodes of glaciation. In 1881, Chamberlin was appointed chief geologist of the Pleistocene division of the US Geological Survey. Six years later, while retaining that position, he also became president of the University of Wisconsin. In 1892, still with the USGS, he moved to the new University of Chicago to assemble a geology department. In 1894, he joined Robert Peary's expedition to Greenland. Chamberlin authored a well-regarded textbook of geology, became president of several scientific societies and recipient of honors, including election to the National Academy of Sciences. Altogether, he had substantial bona fides as his work became increasingly speculative, including views on the formation of the solar system and the causes of repeated episodes of glaciation.

Chamberlin's explanation for multiple ice ages was based on the prior work of Tyndall and Arrhenius, though with emphasis on the latter's view that fluctuations in atmospheric CO_2 are key. Chamberlin was perhaps the first to stress the importance of interactions among the land, the sea, and the air, all of them huge reservoirs of carbon from both organic and inorganic sources. He proposed feedbacks whereby these reservoirs sometimes drained carbon from the air and at other times added it, producing falling or rising temperatures

It was obvious that there must have been relatively large variations of CO_2 in ancient atmospheres, in view of the enormous coal seams of the Paleozoic era, clear evidence of former ages when immense quantities of carbon once in living plants were sequestered under the ground. There were also limestones and other carbonate rocks, and the ocean itself, as repositories of carbon. Considering the small portion of the atmosphere that is CO_2 (today put at 0.04 percent by weight), it would not have taken much change in absolute quantity to double or halve its concentration.

Chamberlin sketched his views in an 1897 article in *The Journal of Geology*, proposing a mechanism whereby CO_2 in the atmosphere would cyclically rise and fall, explaining the alteration of colder glacial with warmer interglacial epochs. He asserted that cold water absorbs more CO_2 than warm water. (Similarly, CO_2 remains dissolved in a refrigerated Coke but bubbles out of solution as the Coke warms.) As atmospheric CO_2 declines, air and sea temperatures also decline, and the capacity of the cooling oceans to take up CO_2 increases, in turn decreasing air temperature further. Also, with increased cold, the process of organic decay becomes less active, so the carbon of dead organisms remains locked up, further depleting the atmosphere of carbon, and thus the epoch of cold is begun.

To account for returning warmth, Chamberlin noted that the spread of glaciation would cover the main crystalline rocks that would otherwise take up carbon from the atmosphere through weatherization. Without this removal, atmospheric carbon increases, and temperature warms. With returning warmth, the ocean gives up its CO_2 more freely, and accumulated organic products decay, both processes adding carbon to the air, and thus warmth accelerates, hastening interglacial mildness.

Today's readers may debate whether or not Chamberlin's speculations would have been published had he not been the journal's editor, but in any case, his prominent pen elevated CO_2 among American geologists as an important driver of climate change, at least for a while. After an initial flourish of attention, critics objected that CO_2 alone could not have so strong an influence on climate. Chamberlin later amended his theory, giving more importance to water vapor, but still support fell away. Toward the end of his life, Chamberlin wrote to a correspondent of his folly in following Arrhenius's emphasis on CO_2: "I greatly regret that I was among the early victims of Arrhenius' error" (quoted in Fleming 1998: 90).

Too bad Chamberlin did not live to see his theory revised and revived. Looking back at Figure 6.3, showing glacial-interglacial cycles over the past 400,000 years, measured from the Vostok ice core, we see peaks and declines in temperature nearly matched, actually slightly lagged, by peaks and declines in atmospheric carbon dioxide (also by methane, not shown). This increase may be explained, at least partly, by changing absorption and release of CO_2 from the ocean around Antarctica, responding to change in ocean temperature (Petit et al. 1999). Thus, changes in the concentration of CO_2 (and other greenhouse gases) in the atmosphere account for some of the worldwide change in temperature that orbital cycles alone do not explain.

The cycles that Milankovitch identified have occurred for hundreds of millions of years. Why, then, did glacial cycles of the Pleistocene not begin until 2.6 million years ago? CO_2 again suggests an answer. Apparently, prior to 2.6 million years ago, levels of atmospheric CO_2 were too high, making Earth too warm to form large ice sheets even when orbits would correspond to those events. CO_2 in the atmosphere became lower by 2.6 million years ago, bringing Earth's temperature closer to freezing, so that orbital changes could periodically shift temperature above or below the ice point.

Changing concentration of CO_2 in the atmosphere may also explain the mid-Pleistocene Transition Point, when the glacial period changed from 41,000 years to 100,000 years. Computer simulations of ancient climate show that a gradual lowering of CO_2 along with erosion of regolith, the layer of unconsolidated rocky material covering bedrock, could have produced this change in periodicity (Willeit et al. 2019).

Sources and Sinks of Carbon Dioxide

But apart from humans burning fossil fuels, how does the atmospheric concentration of CO_2 change? Volcanic eruptions are one contributing source. The oceans and land provide large sinks that remove CO_2 from the air. Atmospheric temperature depends on the balance between sources and sinks. When addition of CO_2 to the air exceeds that taken out, temperature rises. When subtraction of CO_2 exceeds what is added, temperature drops.

Especially important on million-year time scales is the so-called carbonate-silicate cycle, which moves CO_2 from the air to the land to the ocean. Silicate rocks like granite, composed largely of silicon and oxygen, make up most of Earth's continental crust. Carbonate rocks like limestone, made primarily of carbon and calcium, are sedimentary, sometimes composed of fossil seashells. The carbonate-silicate cycle, as its name suggests, describes the interaction of these two kinds of rock (Walker et al. 1981).

In the air, gaseous CO_2 combines with water vapor to form carbonic acid. This falls on land as acid rain, weathering silicate rock surfaces, dissolving their minerals and carrying them to the sea, where they are used by marine organisms like foraminifera. Thus, silicate rocks are transformed into carbonate seashells. When marine organisms die, they settle to the sea floor and are buried in deep sediments. In the process, CO_2 has been removed from the atmosphere and locked in the seafloor. Eventually, when the seafloor is subducted into the mantle, its carbonites are heated and recombine with silicate minerals, a reaction producing CO_2. Volcanos release gaseous CO_2 back into the atmosphere, completing the cycle. Apart from orbital mechanics, the concentration of carbon dioxide in the atmosphere is key to understanding long-term climate change.**

Variations in water vapor, an effective greenhouse gas, do not contribute much to present-day warming because water precipitates out

** There are important greenhouse gases in addition to CO_2 and water vapor, identified by Tyndall, especially methane (CH_4), the major component in natural gas, and nitrous oxide (N_2O), created during the combustion of fossil fuels. Also there are synthetic fluorinated gases, emitted from various industrial processes, usually in small quantities but because of their high warming potential included among the important GHGs.

of the atmosphere within days. Also, cloud formation from water vapor reflects away sunlight, thereby reducing greenhouse warming. This process is very like Thomas Chrowder Chamberlin's speculative explanation of more than a century ago for glacial periods. Recall that in retrospect he "greatly regretted" his emphasis on CO_2. But his ideas are again in vogue, perhaps because of so much interest today in rising levels of carbon dioxide. A recent article in the prominent journal *Science*, advancing an explanation for ancient ice ages, gave a nod to Chamberlin by including him in its opening citations, an unusual tribute for a century-old scientific paper.

Authors of the *Science* article hypothesize that when ancient continents collided with and accreted volcanic arc continents, they must have exposed previously submerged silicate rocks to weathering. This weathering, strongest in the wet tropics, would have increased the sequestration of atmospheric CO_2, lowering global temperature. In support of their hypothesis, the authors report a strong correlation between the extent of glaciation at different times during the Phanerozoic, and surface accretions from such arc-continent collisions (Hartmann 2019; Macdonald et al. 2019). Chamberlin in his coffin must have smiled.

Human-Caused Global Warming

We have lived for over 10,000 years in a relatively warm and stable interglacial period. One would expect, from Milankovitch cycles and the natural waxing and waning of atmospheric CO_2, an inevitable cooling and regrowth of ice sheets, but this may never happen.

Which GHG has the greatest warming effect on the atmosphere? One ton of methane absorbs much more energy than one ton of CO_2, but methane emitted today remains in the atmosphere only about a decade, while CO_2 emitted today will remain in the air for centuries. The combined effect of shorter lifetime and higher energy absorption is considered in the calculation of a *Global Warming Potential* (GWP), which is defined as 1 for CO_2. The time period usually used in this calculation is 100 years. This gives methane a GWP of 28–36, meaning that a ton of methane is roughly thirty times more powerful in warming the atmosphere than a ton of CO_2. Nitrous oxide has a GWP of 265–298 and remains in the atmosphere for over a century. The synthetic fluorinated gases can have GWPs exceed a thousand.

The values of GWP depends on the time period assumed, usually 100 years. A 20-year GWP is sometimes used, essentially giving priority to gases with shorter lifetimes in the atmosphere. For example, the 20–year GWP for methane goes up to 84–87, whereas that for CO_2 remains, by definition, 1. A gas's GWP must be considered in conjunction with the amount of it in the atmosphere. For the United States in 2017, GHG emissions were 82 percent carbon dioxide, 10 percent methane, 6 percent nitrous oxide, and 3 percent fluorinated gases (www.epa.gov). Carbon dioxide is generally considered the major greenhouse gas driving global warming.

Since the Industrial Revolution began in eighteenth-century Britain, humans have burned always increasing amounts of fossil fuel, first coal, then adding petroleum, then adding natural gas. Anthropogenic carbon dioxide, and other greenhouse gases, are discernibly warming the atmosphere and the oceans. This is causing diverse effects, including rising sea levels and the periodic or permanent inundation of low-lying land. Much of Bangladesh, for example, could be under water by the end of this century. If it were a rich nation, it could build seawalls to hold back the water, as Holland has done for decades.

The discoverers of "hot house gases" did not foresee them causing a great problem. Recall Arrhenius's opinion that humans would enjoy any increased warmth as relief from an otherwise cooling climate. During the 1950s, scientists were concerned about pollution from burning fossil fuels, but not about global warming. The oceans were so great a reservoir of CO_2, far more than air, that it was assumed any excess put into the atmosphere would be absorbed by the sea (Weart 2008).

Oceanographer Roger Revelle (1909–1991), director of the Scripps Institution of Oceanography in California, was one of the few who wondered if this assumption was correct (Revelle and Suess 1957). Revelle recruited Charles Keeling (1928–2005), a postdoc at Caltech working on CO_2 measurement, to continue his work at Scripps. Obtaining funding, Keeling began monitoring atmospheric CO_2 at the observatory atop the volcano of Mauna Loa in Hawaii, a site considered free of extraneous carbon contamination. His monitoring program, which continues today, soon showed that carbon was increasing in the atmosphere and continues to do so (Figure 7.1). At that point, it was unclear if the increase was causing a detectable warming of Earth's atmosphere.

Since the nineteenth century at least, individuals have recorded more or less continuous records of local temperatures, most frequently in Europe and the United States. Following no consistent standard, these were difficult to aggregate into a global trend, though the attempt was occasionally made. The effort intensified in the late twentieth century, producing what appeared to be a warming trend, however there were difficulties with this interpretation. Most measuring stations were in the industrial nations of the West, few at sea or from the Southern Hemisphere. Another problem was the "heat island" effect of cites, i.e., heat retention by concrete, brick and pavement, and heat

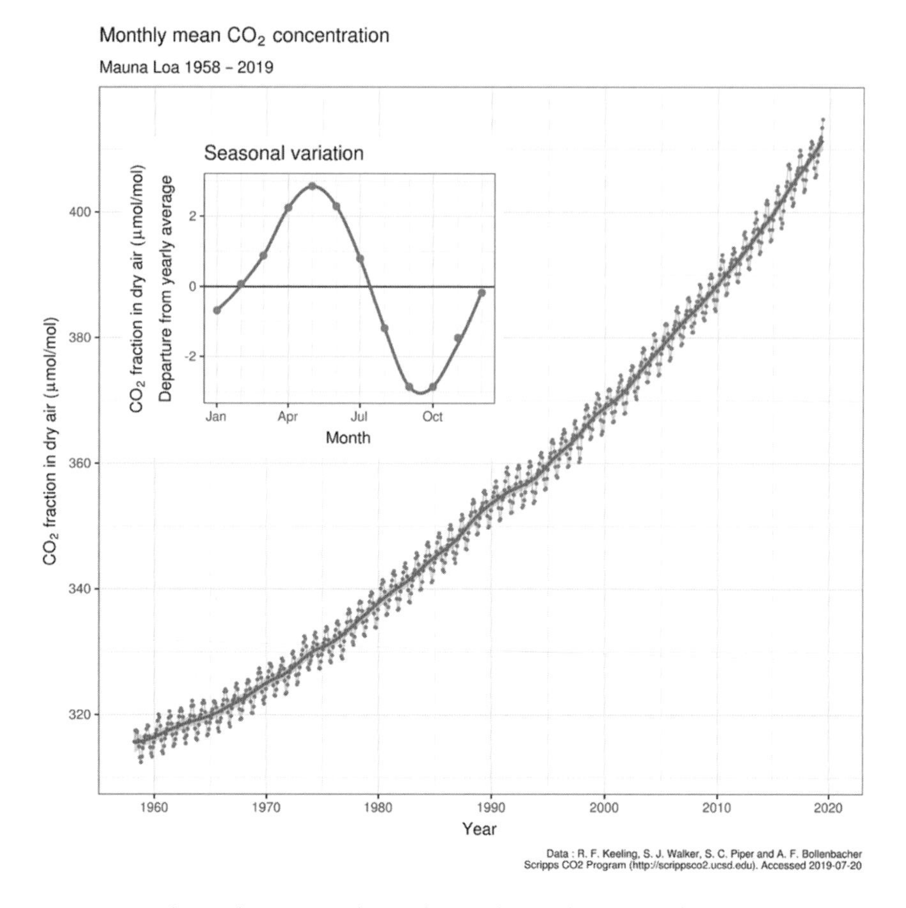

Figure 7.1 The Keeling Curve shows the yearly trend in atmospheric CO_2 as measured directly at Mauna Loa, Hawaii, from 1958 to 2019. There are annual squiggles because foliage, most in the Northern Hemisphere, grows leaves around April, thus absorbing CO_2, and sheds them around October.
Source: NOAA

generation by furnaces and engines, make cities warmer than the countryside. With passing years, thermometer sites in rural areas were approached if not surrounded by growing cities and suburbs. Possibly the apparent rise in temperature was an artifact of heat islands moving closer to measuring sites. Over time, corrections were applied and the distribution of measuring stations was widened, providing reasonably valid trends. Eventually satellites provided more trustworthy global temperatures but for a shorter duration. There are now multiple reliable temperature trend lines, as in Figure 7.2.

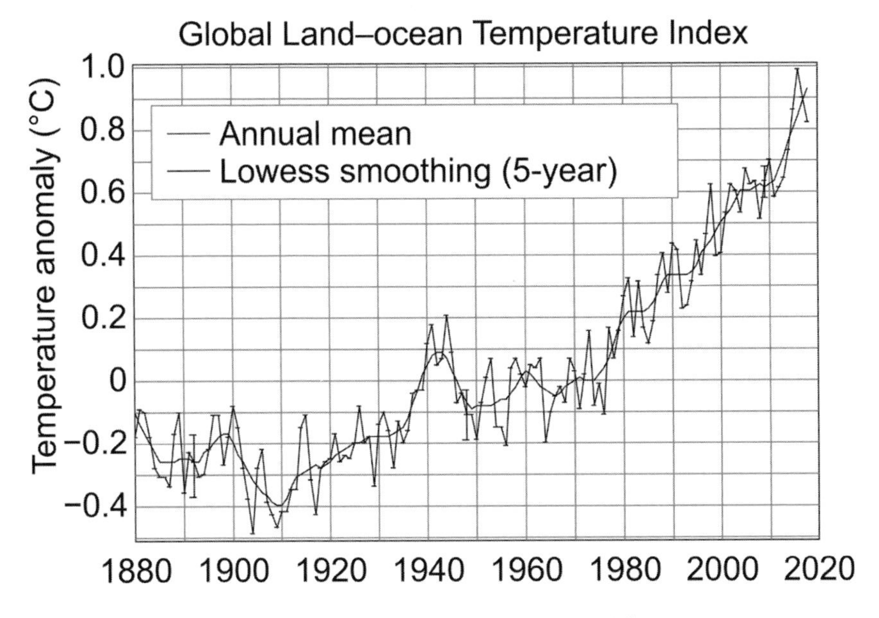

Figure 7.2 Land-ocean temperature index, 1880 to 2019. The temperature anomaly is each year's departure from the average temperature during the base period 1951–1980. The jagged line is the global annual mean, and the smoother line shows five-year averages.
Source: NASA Goddard Institute for Space Studies

Looking back to the 1970s, a major problem for climate theorists was that since about 1940, global temperature had leveled off or even declined! Some reasonable scientists concluded that rising CO_2 had been insufficient to affect global climate. No one doubted that greenhouse gases trapped heat in the atmosphere, but the magnitude of effect – whether it made much difference – was very much in doubt.

The uncertainty was shown in *The Global 2000 Report*, commissioned in 1977 by then-President Jimmy Carter but not released until 1980. It was a forward-looking study intended to anticipate stresses involving population, resources, and environment. The authors, aware of concern about rising CO_2 in the atmosphere and its potential for causing warming, addressed these in their chapter 17, "Climate," first noting the great deal of disagreement among climatologists over what changes to expect. Interviewing 24 climate experts about likely warming versus cooling by the year 2000, they found opinions nearly split.

Ups and Downs of Public Concern

Global warming was little noted by the public, news media, or government before 1988. In summer of that year, during an extreme heat wave and drought in the eastern United States, NASA climate scientist James Hansen testified before Congress, as he had on three other occasions to little attention, but this time the hearing room was full of reporters cued to expect an important story related to the drought. That evening NBC television news showed Hansen's statement that global warming is happening now; the next evening's broadcast connected it with the drought. (Hansen later agreed that long-term climate change cannot be blamed for one season's anomalous weather.)

Major newspapers highlighted the story.[1] It was on page one of the next day's *New York Times*. *Newsweek*'s cover of July 11, "The Greenhouse Effect," was pegged on the drought and Hansen's assertions. News reports during the summer of 1988 about hypodermic needle-polluted beaches on the East Coast (associated with the AIDS epidemic) and a massive forest fire in Yellowstone National Park amplified the image of intense heat.

Skyrocketing coverage of global warming carried with it stories about man-made fires during the Amazon dry season, used to clear sections of rainforest for planting. Reporters flew to the Amazon to film the conflagration. (Amazon fires were more extensive the prior year but went unreported in the United States.) These were juxtaposed against the huge Yellowstone fire. Appreciating the press attention in 1988, biodiversity activist E. O. Wilson commented to me, "It's a pity Yellowstone could only burn once."

The clustering together of global environmental problems was by then common and received news coverage around the world. George H. W. Bush, during his 1988 campaign, announced that he would be an environmental president. *National Geographic* magazine, for its final cover of 1988, featured a hologram of a crystalline "Fragile Earth" being pierced by a bullet. (The back of the magazine carried a hologram of a McDonald's restaurant, sponsor of this extraordinary cover.) *Time* magazine, instead of naming as usual a Person of the Year for 1988, featured "Endangered Earth" as its Planet of the Year. The Exxon

[1] For documentation of media coverage and public opinion polling, see Mazur (2018).

Valdez oil spill of March 1989 drove environmental attention still higher. This crescendo of media coverage and public concern reached its climax on Earth Day 1990, the most widely celebrated ever.

By 1992, US press coverage was waning, even as the presidency passed to Bill Clinton and his environmentalist vice president, Al Gore. The sudden outbreak of the first Gulf War of 1991, despite publicizing oil well fires started by Iraqi forces, seemed to have broken the flow of stories on the global environment. The collapse of the Soviet Union that year was a colossal event pre-empting news space. Global temperatures fell in 1991 and 1992, probably because of sulfur aerosols produced by the Pinatubo volcanic eruptions in the Philippines. Perhaps these factors contributed to the expiration of the "Endangered Earth" as a news story. With declining media covered, public concern turned elsewhere.

News coverage rose again in 2006–2008, higher than before, initially in the United States, then worldwide. Certainly Al Gore with his slideshow, "An Inconvenient Truth," was one factor in the revival of media attention. There were other contributors, perhaps including the objectively rising temperature of Earth's atmosphere. US gasoline prices were higher in 2006–2007 (in constant dollars) than they had been since 1980 in the aftermath of the revolution in Iran. Hurricane Katrina in August 2005 and the prolonged misery of New Orleans were fresh in mind as warnings of more intense storms were linked to global warming. (In fact, the 2006 and 2007 hurricane seasons were unexpectedly mild.)

Al Gore had re-emerged with his slideshow after his hairbreadth loss to George W. Bush in the 2000 presidential election. To liberals, he now seemed warmer and wiser than he had been as a wooden presidential candidate. Some conservative commentators likened him to a snake oil salesman, spewing a lot of hooey. What had already been a Democratic/Republican split over the reality of global warming (and environmentalism generally) became more polarized. Then the Great Recession of 2008, precipitated by the sudden collapse of Lehman Brothers in September of that year, turned public and journalistic concerns about warming into the background.

In December 2015 in Paris, global leaders celebrated the landmark United Nations agreement on reducing greenhouse gases, approved by nearly every nation in the world. The accord's aim is to keep global mean surface temperature well below 2°C above the preindustrial mean and, if possible, to limit the increase to 1.5°C. In the United States, seven of the Republican contenders for the presidency publicly doubted the reality of man-made climate change, while five did not deny the science but opposed regulation because of its costs to the economy (Andrews and Kaplan 2015).

US news media turned again to climate, perhaps spurred by the startling election to the American presidency of Donald Trump, who would soon announce the withdrawal of his nation from the Paris Accord, and would reverse climate-protecting initiatives signed by the prior (Obama) administration. Media coverage reached unprecedented heights internationally, carrying warnings that if corrective actions were not soon taken, global warming would become unstoppable. In the United States, some progressive Democrats proposed a "Green New Deal" to virtually eliminate fossil fuels by 2050, a highly aspirational if practically unattainable goal though viewed sympathetically by President Joseph Biden.

The Intergovernmental Panel on Climate Change

Back in 1988, when global warming first became a public issue, the United Nations and the World Meteorological Organization created the Intergovernmental Panel on Climate (IPCC) to assess the evolving scientific understanding of climate change, its impacts, and the potential for mitigation (www.ipcc.ch/). The IPCC would assess the probable results of various policy options; it could not recommend specific actions. The intent was to use scientists for what they do well, which is make scientific assessments, but not to propose government policy, for which they have no special wisdom.

Issuing its first assessment in 1990, the IPCC continues to release updated reports about twice per decade. These assessments are extraordinarily comprehensive and transparent. For each report, there is first a round of expert review, including comments from thousands of scientists. Lead authors for each chapter of the report must consider and respond to all comments and make appropriate revisions. The revised draft is reviewed again, this times by representatives of all participating UN member nations as well as climate experts. The final draft of the all-important summary for policymakers is agreed upon, word for word, in a plenary meeting of government delegations, an opportunity for political input. At the 1995 plenary in Madrid, for example, there was an intense debate between scientists authoring the report and the delegates from Saudi Arabia who said the word "appreciable" was too strong in the proposed statement: "The balance of evidence suggests that there is an appreciable human influence on climate." After two days of argument, a compromise was reached by replacing *appreciable* with *discernible*, and that is how the summary was published (Mann 2012).

As the pace of research increased, results became clearer. Temperature trends were verified, and proxy analyses of global climate extended backward in time. It was recognized that industrial activity during World War II and the postwar economic boom added sulfate aerosols to the atmosphere, reflecting incoming sunshine away from Earth, hence the cooling trend of those years until pollution controls were established. Then warming returned. While uncertainties remain, the evidence increasingly showed that atmospheric warming was at least partly due to humans burning fossil fuels. The IPCC's fifth assessment report, released in 2014, stated, "Human influence on the climate system is clear, and recent anthropogenic emissions of greenhouse gases are the highest in history. Recent climate changes have had widespread impacts on human and natural systems" (IPCC 2014). Among atmospheric scientists, excepting a few contrarians, there is nearly consensual agreement today. The IPCC and Al Gore were jointly awarded the Nobel Peace Prize in 2007.

At this writing, the most recent IPCC assessment is the first part of its sixth report (AR6), released in 2021. It concluded in part:

> Observed increases in well-mixed greenhouse gas (GHG) concentrations since around 1750 are unequivocally caused by human activities. Since 2011 (measurements reported in AR5), concentrations have continued to increase in the atmosphere...
>
> Each of the last four decades has been successively warmer than any decade that preceded it since 1850...
>
> It is likely that well-mixed GHGs contributed a warming of 1.0°C to 2.0°C, other human drivers (principally aerosols) contributed a cooling of 0.0°C to 0.8°C, natural drivers changed global surface temperature by –0.1°C to 0.1°C, and internal variability changed it by –0.2°C to 0.2°C. It is very likely that well-mixed GHGs were the main driver of tropospheric warming since 1979...
>
> Globally averaged precipitation over land has likely increased since 1950, with a faster rate of increase since the 1980s (medium confidence). It is likely that human influence contributed to the pattern of observed precipitation changes since the mid-20th century, and extremely likely that human influence contributed to the pattern of observed changes in near-surface ocean salinity.

> Human influence is very likely the main driver of the global retreat of glaciers since the 1990s and the decrease in Arctic sea ice area between 1979–1988 and 2010–2019... It is very likely that human influence has contributed to the observed surface melting of the Greenland Ice Sheet over the past two decades, but there is only limited evidence, with medium agreement, of human influence on the Antarctic Ice Sheet mass loss.
>
> It is virtually certain that the global upper ocean (0–700 m) has warmed since the 1970s and extremely likely that human influence is the main driver...
>
> Global mean sea level increased by 0.20 [0.15 to 0.25] m between 1901 and 2018. The average rate of sea level rise was 1.3 [0.6 to 2.1] mm/yr between 1901 and 1971, increasing to 1.9 [0.8 to 2.9] mm/yr between 1971 and 2006, and further increasing to 3.7 [3.2 to 4.2] mm/yr between 2006 and 2018 (high confidence). Human influence was very likely the main driver of these increases since at least 1971...
>
> In 2019, atmospheric CO concentrations were higher than at any time in at least 2 million years (high confidence), and concentrations of CH_4 and N_2O were higher than at any time in at least 800,000 years (very high confidence). Since 1750, increases in CO_2 (47%) and CH_4 (156%) concentrations far exceed, and increases in N_2O (23%) are similar to, the natural multi-millennial changes between glacial and interglacial periods over at least the past 800,000 years (very high confidence).
>
> It is virtually certain that hot extremes (including heatwaves) have become more frequent and more intense across most land regions since the 1950s, while cold extremes (including cold waves) have become less frequent and less severe, with high confidence that human-induced climate change is the main driver of these changes...
>
> Many changes due to past and future greenhouse gas emissions are irreversible for centuries to millennia, especially changes in the ocean, ice sheets and global sea level...

Persistence of Controversy

Perhaps, naively, the IPCC founders thought they could keep the objective science of climate change above the fray, leaving the battling over what to do about it to politicians and lobbyists. They were wrong. As the scientific evidence of human-driven warming grows stronger, contrarians continue to attack the science, sometimes on minor points, sometimes claiming conspiratorial actions. I have at hand a 3-pound volume, recently arrived at my office from the Heartland Institute, a politically conservative and libertarian think tank that works against the consensus science on climate change and other issues including harm from secondhand cigarette smoke. This volume takes aim at the credibility of the IPCC, documenting contrary findings. I was surprised to find in its Forward the almost-concessionary statement that the editors "find that *while climate change is occurring and a human impact on climate is likely*, there is no consensus on the size of that impact relative to natural variability, the net benefits or costs of the impacts of climate change, or whether future climate trends can be predicted with sufficient confidence to guide public policies today" (Bezdek et al. 2019: vii, my italics).

The central paradox of many public controversies that have at their core a dispute over scientific, medical, or engineering facts, is that resolution of these technical questions has little effect on the controversy. Examples that may be familiar to readers are the insistence of those resisting vaccinations for children that these cause autism, despite abundant evidence to the contrary, or the demands of some religious fundamentalists that public schools teach some version of biblical creationism alongside our modern understanding of the evolution of life. Granted that most factual disputes are not so definitively settled as evolution and the billions-of-years age of Earth, the weight of evidence for human-caused global warming is substantial.

8 WHY DOES CLIMATE CHANGE?

Continental Drift and Ocean Currents

Charles Lyell speculated in the 1834 edition of his *Principles of Geology* that long-term variations in climate would occur under redistributions of land and ocean. He was not addressing the cause of ice ages, of which he was not then aware, but prior changes in climate, both hotter and colder than in his day, as inferred from the fossil record. He illustrated his speculation with two contrasting global maps, the first showing a clustering of continents in equatorial latitudes, surrounded by ocean. The second map showed continental land masses divided between the two poles, with ocean covering the equatorial latitudes (Figure 8.1). He was not anticipating continental drift but envisioning land and sea trading places though long-term depression of land and elevation of seafloor.

Lyell knew that ocean currents such as the Gulf Stream carried heat from the equator to the North Atlantic, accounting for Britain's warmer climate than similar latitudes in North America, and that air currents did the same. He knew that land, especially high mountains, was colder than the ocean. He was aware of the albedo effect, that great expanses of white snow and ice would reflect more sunlight than unfrozen land or sea, thus reducing the overall heat absorbed from the sun. All these contributed to his speculations. In his top configuration all land is near the equator, receiving maximum sunlight. Air and ocean currents are unimpeded in transferring heat from equator to the poles. There would be little ice to reflect away sunlight. In his second configuration, all land masses are around the poles where they would receive less sunlight than if near the equator, currents from the equator cannot

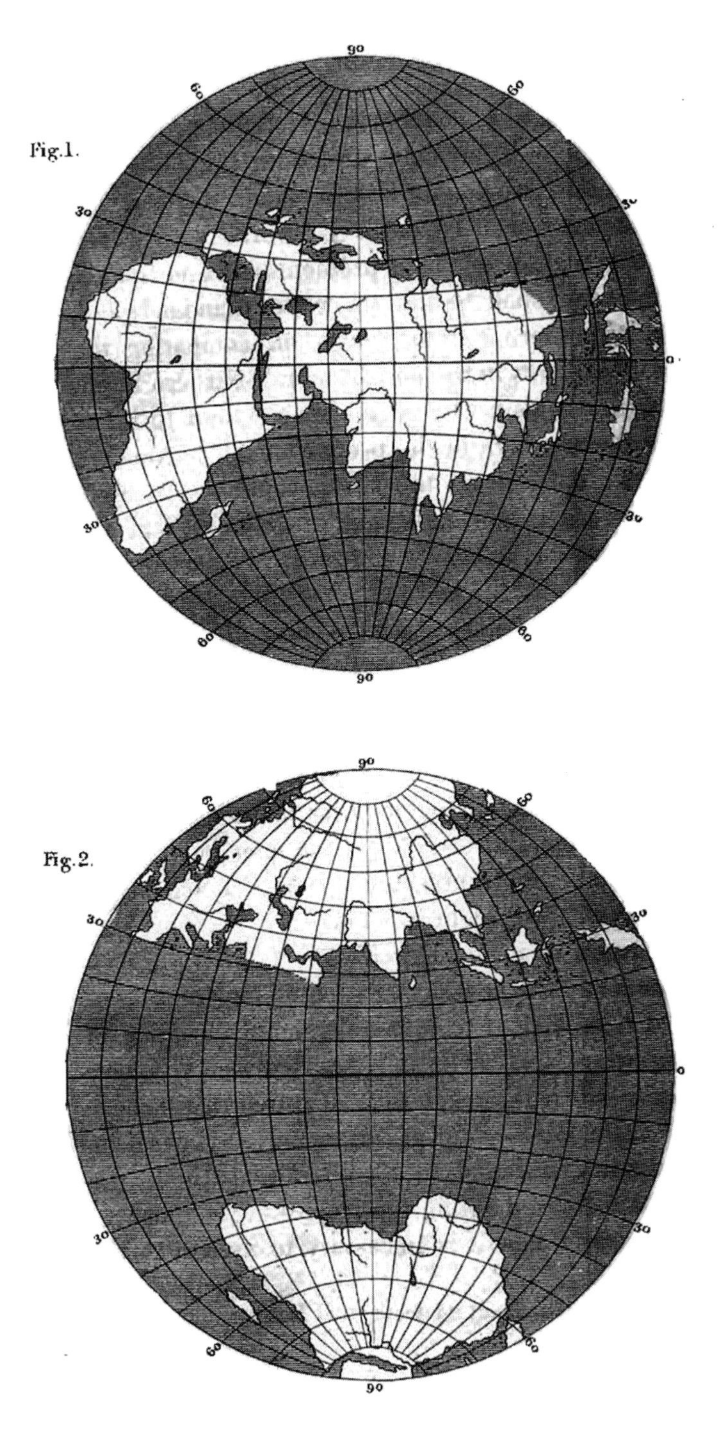

Figure 8.1 Lyell's contrasting distributions of land on Earth's surface.
Source: Lyell 1834, Vol. one: 188

reach the polar regions to deliver their warmth, and blankets of snow and ice at the two poles would maximize albedo, reflecting away sunlight. Thus, these contrasting configurations produced extremes of hot and cold global climates.

No geologist prior to the twentieth century believed that continental configurations were in the past different from the current one. Today we know that all major land masses were at some time near the equator, and if this occurred within the last half a billion years, their indigenous life evolved in a warm environment.

Amadeus Grabau (1870–1946) was a professor fired by Columbia University who then spent the rest of his life at Peking University and is still honored there for bringing Western earth science to China. He and his Chinese colleagues were among the earliest exponents of continental drift (Mazur 2004). Remarkably advanced in his reconstructions of former continental positions, Grabau showed through fossil distributions that Europe and America had once been connected, before today's Atlantic Ocean opened between them. Ironically, Grabau used continental drift to argue *against* the idea that alterations in Earth's climate, between warm periods and cold, caused successive glacial and interglacial epochs. He proposed instead that whatever land surface was in a polar region would have glaciers, simply because the poles have lower temperature than middle latitudes. When that same land surface eventually moved toward the equator, its ice would melt and its climate turn tropical. Antarctica, which indeed was once tropical, as known from its fossils, is presently ice covered because it now (temporary) sits at the South Pole.

It was a clever, peculiarly time-bound argument, but it's wrong. Neither Lyell nor Agassiz could have made such a mistake because they had no notion that the continents move. Grabau did, being among the first adherents to Wegener's view, decades before its general acceptance. If Grabau had lived two decades longer and so had known from radiocarbon dates that the last great Ice Age was too recent to be affected by shifting positions of Europe or North America, he would not have made this blunder. His problem was one of having advanced knowledge, but not advanced enough. Today we know that the continents barely differed from their present positions during the Pleistocene's 2.6 million years of intermittent glacial and interglacial periods.

It is unquestionably correct that Antarctica would not now be glaciated if it were located in temperate latitudes, as it once was.

Global position alone does not fully explain glaciation, but it is relevant, as is the configuration of land masses and the constraints they place on ocean and air currents. In times past, there was no land at either pole, so ocean currents could flow unimpeded, carrying equatorial warmth poleward, leaving these areas of the sea unfrozen. Today the continent of Antarctica prevents this warming effect. There is no land at the North Pole, but a nearly landlocked Arctic Ocean, almost surrounded by land masses, has the same effect, making it difficult for warm water to flow northward. The Gulf Stream, flowing north, would warm Arctic waters if it were not deflected by the eastern bulge of Canada and by Greenland, flowing to the British coast. If the Gulf Stream stopped flowing, the British climate would become considerably cooler, its winters like those of Canada at the same latitude.

The Global Conveyor Belt

The Gulf Stream originates in the warm Gulf of Mexico. Driven by prevailing winds, it flows around Florida and then northward along the southeastern coast of the United States. Roughly 100 km wide, the volume of water in the Gulf Stream is far greater than the combined flow of all rivers that empty into the Atlantic.

The Gulf Stream became known to Europeans from Ponce de León's 1512 exploration of the New World. The explorer reported that his ship could not make headway under sail when traveling against the strong flow. Subsequently Spanish ships learned to use the Gulf Stream when returning from the Caribbean back to Spain, sailing in the direction of the current.

Benjamin Franklin, interested in all things, charted the flow of the Gulf Stream farther up the American coast. He saw that its northeastern flow was turned by protruding Canada across the Atlantic toward the European coast. Though Franklin did not know it, the stream then turns north again, toward Greenland.

The Gulf Stream is now recognized as part of a huge pattern of ocean circulation called "the Global Conveyer Belt," or "the Thermohaline Circulation" (Figure 8.2), named for the temperature and salinity that affect water density and drive the circulation (Wunsch 2002). It has the overall effect of carrying cold water from the polar regions toward the equator, and warm water from the equator toward the poles.

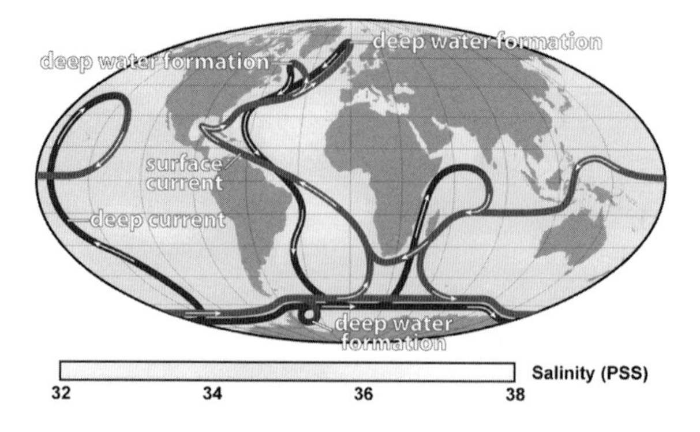

Figure 8.2 Global Conveyor Belt, or Thermohaline Circulation. Arrows indicate the changing direction of flow, sometimes on the surface, sometimes below. Water becomes colder and denser as it approaches the poles, causing it to sink into the deep ocean, driving the circulation. *Source*: NOAA

With far earlier continental configurations, warm water could reach all the way to the poles so they were ice free, as Lyell imagined.

When ice forms in the polar regions, the salt is left out, making the surrounding seawater saltier as well as colder, thus increasing its density so it sinks. Surface water is pulled in to replace the sinking water, which in turn becomes cold and salty enough to sink. This initiates the deep-ocean current driving the global conveyer belt.

Very deep cold water from the North Atlantic moves south, between the continents, across the equator, past the tips of Africa and South America, then traveling around the edge of Antarctic. There it cools and sinks again, in effect "recharging" the conveyor belt. Continuing around Antarctica, the stream splits in two, both branches turning northward. One branch moves into the Indian Ocean, the other into the Pacific Ocean. Both warm as they approach the tropics, becoming less dense, so they rise to the surface (upwelling). Blocked by land, they loop back southward and westward to the South Atlantic, eventually returning to the North Atlantic, where the cycle begins again (Broecker 2010).

The conveyor moves much slower than wind-driven or tidal currents. It is estimated that one unit of ocean water takes about 1,000 years to complete the full journey. The total flow is enormous, about 100 times that of the Amazon River (www.oceanservice.noaa.gov).

If the conveyor is disrupted by the intrusion of freshwater into the North Atlantic, this dilutes the salt and blocks formation of sea ice, disrupting the sinking of cold, salty water, thus slowing down or turning off the global conveyor.

Ice and seabed cores show that during the Pleistocene, there were episodes of very rapid climate change, several degrees in a decade. There is speculation that this occurred because of interruption in the global conveyer belt, ceasing its activity of redistributing heat around the planet. One proposal assumes that a giant reservoir of meltwater accumulated behind an ice dam on Canada's Laurentide ice sheet. Then the dam broke, inundating the North Atlantic with freshwater that diluted the dense salty seawater so it no longer sank into the ocean, shutting down the conveyer. This is no evidence this scenario actually occurred, but the cataclysmic breaching of ice dams is a known phenomenon.

The Columbia River in Montana once had ice dams up to 600 meters high, impounding melt water to form glacial Lake Missoula, covering 7,800 km^2 and containing half the volume of modern Lake Michigan. About 14,000 years ago, an ice dam was breached, releasing torrents of water downstream to create the Channeled Scablands of eastern Washington State. Up to 40 cycles of damming and breaching have been reported, making an average life span of a lake and consequent outburst flood about every 50 years. The earliest Americans may have witnessed these gigantic floods (Blacey and Ranney 2018).

Moving Continents

Alfred Wegener published the third edition of his book on drift, *The Origin of Continents and Oceans*, in 1922, and it was this version that captured the attention of Amadeus Grabau and his Chinese colleagues at Peking University. The geological picture then emerging showed the Paleozoic earth with a single supercontinent (Pangaea). It subsequently broke apart, and the continental blocks moved slowly to their present positions while opening up the Atlantic Ocean. Wegener integrated physical and biological evidence as supporting evidence. The best-known physical argument is that the east coast of South American and the west coast of Africa fit together like pieces of a crude jigsaw puzzle. The fit improves when submerged continental shelves are matched. If the continents are thus reattached, nearly identical rock formations on both sides of the Atlantic become contiguous.

The primary biological argument emerged in the nineteenth century as colonizing Europeans, engrossed with natural history, collected plants and animals from distant lands. They found similar species of weeds, mammals, snails, and earthworms in widely separated places, with no land connection between them. Charles Darwin was engrossed by this inexplicable dispersion. He tested how long various seeds remained viable in salt water and speculated whether birds might eat spores in one place and then carry them over water for deposition in their droppings, or whether monkeys might float across water on natural rafts of tangled vegetation. Such mechanisms explain dispersion of easily portable species, or across short expanses of water, but they seem implausible for large or fragile species crossing oceans.

Former land bridges, now covered by ocean, became a favored means of explanation. Certainly the shallow Bering Strait was dry when ocean levels were lower. Other postulated land connections were far-fetched, like a supposed continent that once lay between South America and Africa, later sinking into the deep Atlantic. Physicists pointed out that continental blocks, primarily granite, have lower specific gravity than the ocean substrate, primarily basalt. A continental block can no more sink through the ocean floor than an iceberg can sink beneath the ocean's surface.

Continental drift provided an alternate explanation. Any animal that once ranged over Pangaea could have descendants on its subsequently separated pieces. When Wegener, Grabau, and others reconstructed the ancient alignment of continents, what had been disparate locations of specific animals, plants, and fossils came together (Figures 8.3 and 8.4).[1]

Geologists could not see a reasonable mechanism whereby continental fragments could move around Earth's surface, so they almost unanimously rejected continental drift until the 1960s (Newman 1995; Oreskes 1999). Then, in a stunning reversal based wholly on physical evidence from the sea floor, plate tectonics was born.

[1] Even today, continental drift does not solve all these puzzles of dispersal. It remains mysterious how South American monkeys became separated from Old World monkeys (in Africa and Asia). According to modern dating, South America split from Africa one hundred million years ago, well before there were any monkeys at all. After monkeys did evolve in the Old World, there was no land route to South American until the Isthmus of Panama formed less than three million years ago, when monkeys were already well established in South America. Unlikely as it seems, a vegetated raft or rafts carrying monkeys from Africa to South America is the only scenario that plausibly explains how they crossed the Atlantic Ocean (Godinot 2000; de Queiroz 2014).

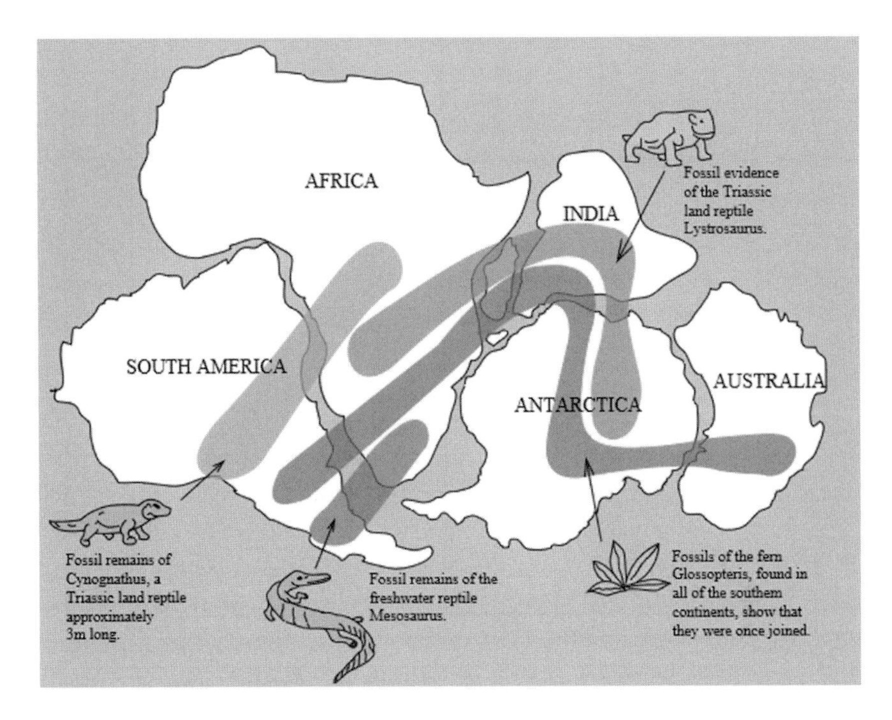

Figure 8.3 Wegener's hypothetical reconstruction of ancient Gondwanaland, with a configuration of continental pieces bringing spatial continuity to what today are disparate distributions of animals, plants, and fossils.
Source: Wikipedia, based on USGS data

The occasional reversal of Earth's magnetic field was key to confirmation. There has been no flip in polarity since compasses were invented. But, as noted in Chapter 5, iron particles in molten lava line up with the direction of the magnetic field at that time and hold their alignment when the lava hardens.

Drift proponents like Grabau argued that the Atlantic Ocean was widening on either side of a rift running north-south along the middle of the ocean basin. They envisioned molten rock from Earth's interior extruding from this mid-Atlantic rift, and then hardening to form new seafloor. If true, magnetism in the new extrusion would polarize in the direction of Earth's magnetic field. After every reversal of the field, newly extruded magna would show a flip in magnetic polarity. Thus the seafloor, expanding outward on either side of the mid-Atlantic rift, should act like a tape recorder, registering reversals of Earth's magnetism.

During the Cold War, the U.S. Defense Department invested in ocean research because here was the arena for submarine warfare.

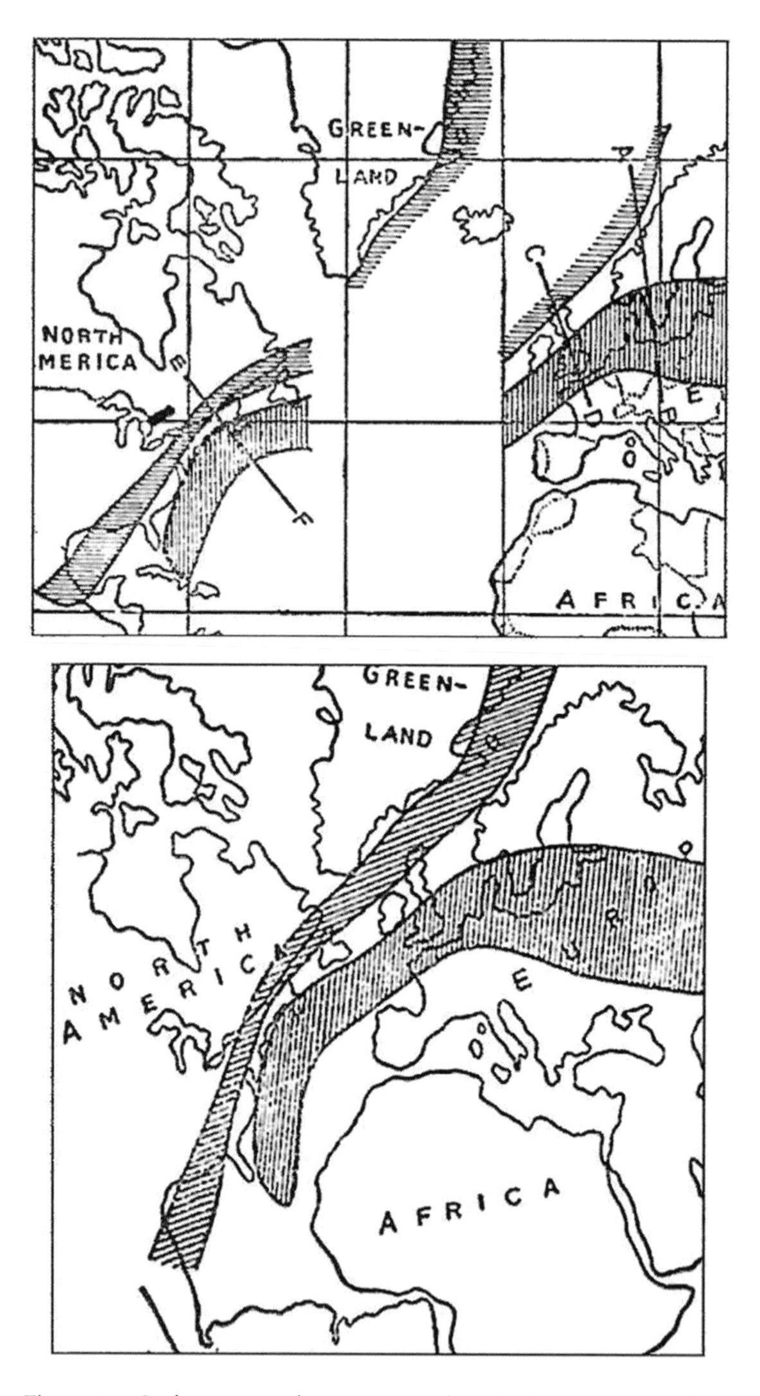

Figure 8.4 Grabau's maps demonstrating that North America and Europe were connected. Top: A modern map shows two distinct trilobite

Cruising ships dragged magnetometers to measure spatial variations in seafloor magnetism. Scientists at several universities, especially Columbia, Princeton, and Cambridge in England, saw in these defense-related efforts a way to test the drift hypothesis. By 1965, they found in the data the expected pattern of alternating magnetic stripes, symmetrical on either side of the mid-Atlantic rift. Seafloor widening was vividly confirmed.

Ancient Ice Ages

During the 2.6 million years of the Pleistocene Ice Age, Earth's land masses were in essentially the same positions they occupy today. So continent drift is irrelevant to explaining repeated intrusions and retreats of ice sheets into northern Eurasia and America during that epoch.

Attempts to trace climate history further back are more uncertain than records from ice cores and marine sediments. However, from the presence of erratic boulders, polished rock, and other physical indicators of glaciation, as well as the location of cold-weather vs. tropical fossils, and growth bands in fossilized coral, one can infer the existence or absence of ice sheets at more ancient times. Additional clues come from isotopic measurements on very-old marine fossils from former seas, now found in dry land. Investigators using such reports estimated the latitudinal extent of continental ice sheets throughout the Phanerozoic eon, the 540 million years since the Cambrian explosion of visible life (e.g., Macdonald et al. 2019). Another approach is to identify times when polar icecaps were present or absent, as was done for the new fossil hall of the Smithsonian Institution's National Museum of Natural History, which shows a "roller coaster" history of climate change over the past 500 million years, illustrating that most often the poles were unfrozen (Voosen 2019).

One reconstruction of Phanerozoic climate, inferred from change in the oxygen isotope ratios as measured in brachiopod fossils (Veizer et al. 1999), is shown in Figure 8.5. Dips in the graph indicate

Figure 8.4 *(cont.)* assemblages (indicated by different hatching), each interrupted by the Atlantic Ocean. Bottom: The fragmented assemblages become contiguous when mapped before the Atlantic Ocean opened. Virtually identical maps were used by geophysicist J. Tuzo Wilson in his 1996 article in *Nature* supporting plate tectonics.
Source: Mazur 2004

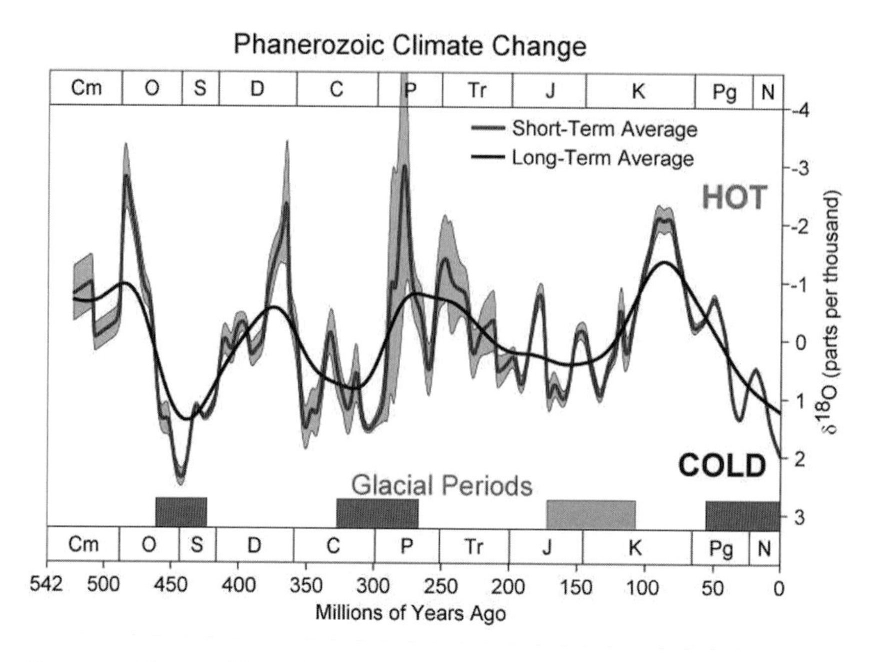

Figure 8.5 The trend line of Phanerozoic climate is based on oxygen isotope ratios in fossil brachiopods. The horizontal bars are based on geological signs of glaciation. The Jurassic-Cretaceous period, plotted in lighter gray, is interpreted as a "cool" period when the configuration of continents at that time may have prevented the formation of large ice sheets.
Source: Dragons flight, Wikipedia Commons

cold periods, or ice ages. This proxy shows four glacial periods in the Phanerozoic, separated by warm periods millions of years long.

The rightmost decline in the graph, following a very warm (K or Cretaceous) period in which dinosaurs lived, shows the post-extinction descent of climate into the *Late Cenozoic Ice Age* (LCIA). Its beginning is dated 34 million years ago when the present Antarctic ice sheet formed. The Pleistocene, the focus of this book, is the tail end of the LCIA, when extensive ice sheets spread over the Northern Hemisphere, sometimes waxing, sometimes waning. We are still in it, though in a mild interglacial episode that has lasted for over 10,000 years.

The 34 million years of the LCIA gave sufficient time for significant changes to the configuration of land. The Isthmus of Panama formed roughly three million years ago, after two seabed plates slowly collided, forming a row of underwater volcanos (Montes et al. 2015; O'Dea et al. 2016). These grew to the surface, becoming islands, eventually consolidating into an isthmus separating the Atlantic and Pacific oceans so they could no longer mix freely. The isthmus also created the Gulf Stream.

About 34 million years ago, the great ocean of the Southern Hemisphere opened, in effect isolating the Antarctic continent over the South Pole. Ocean circulation from the topics could no longer reach the southern polar region, causing it to descend into a deep freeze. Glaciers first formed on the elevated land of East Antarctica, facilitated by a decrease in atmospheric CO_2. Some millions of years later, another ice sheet formed over the lower land of West Antarctica, completely covering in ice what had once been a tropical continent. Today Antarctica contains most of the world's freshwater stored in two large ice sheets, West and East. If these were to melt completely, they would raise sea level by more than 60 meters (Bell and Seroussi 2020). Temperate latitudes were also affected, with the diversification of grasses and the expansion of the grassland ecosystem (Blakey and Ranney 2018).

As preface to the LCIA, the collision of India with Asia was causing the uplift of the Himalayas by 50 million years ago. Associated weatherization is thought to have transferred CO_2 from the air to the sea, weakening greenhouse warming. The colossal Himalayan mountain range altered dominant air currents, reducing the flow of equatorial heat toward the North Pole. These processes combined to gradually lower global temperature from its hot house setting at the time of the dinosaurs' demise, when there were lush forests in the polar regions and ocean temperatures much warmer than today.

The opening of ocean separating Antarctica, South America, and Australia produced by 40 million years ago a powerful current around Antarctica, preventing equatorially warmed water from reaching the large land mass over the South Pole and allowing permanent ice to accumulate around 35 million years ago. Glacial conditions reached Greenland between 20 and 15 million years ago, and the middle latitudes of the Northern Hemisphere around 2.5 million years ago (Woodward 2014).

Apparently, during the Paleozoic era, there were three even earlier periods of extensive glaciation: the Late Ordovician Ice Age (455 to 440 million years ago), the Late Devonian Ice Age (ca. 360 million years ago), and the Carboniferous-Permian Ice Age (335 to 280 million years ago). During the latter, most of Earth's land masses were consolidated into the ancient continent of Pangaea.

Even earlier, before the Cambrian explosion opened the Paleozoic era, there were two periods of extensive ice sheets (National Research Council 1982). The Huron is the earliest identified, dated 2.4–2.1 billion years ago, perhaps before there was any life on Earth. There followed more than a billion years without ice. Then a series of glaciations during the Neoproterozoic Era, between 715 and 547 million years ago, when multicellular life probably existed. (For a timeline and synopsis of ancient ice ages, see https://en.wikipedia.org/wiki/Timeline_of_glaciation)

It has been suggested that pre-Cambrian glaciation was so severe that it caused "Snowball Earth" conditions in which the planet froze over completely. The strongest forms of evidence for this scenario are signs that glaciers formed on land masses that were then at the equator. The reasoning goes that if the warmest of zones was glaciated, the remainder of the planet must have been too. This would have put an icy lock on climate, raising the question of how Earth ever warmed again. One suggestion is that powerful volcanic eruptions released enough carbon dioxide into the atmosphere to raise global temperature above the freezing point. It is difficult to know at present if Snowball Earth is more fantasy or fact (Myrow, Lamb, and Ewing 2018).

9 *ECCE HOMO*

Like the Victorian lady who scoffed at the idea that humans descended from apes but hoped, if true, it would not become widely known, many today do not believe in evolution. Opinion surveys taken biannually from 2006 to 2018 show that nearly half of Americans think it false that human beings developed from earlier species of animals. Seventy-six percent of those belonging to fundamentalist religious denominations do not believe humans evolved, compared to 37 percent of those from moderate or liberal denominations. Sixty-eight percent of those who self-identify as politically conservative do not believe humans evolved, compared to 41 percent of those not conservative, with much overlap between religious fundamentalists and political conservatives.[1] Clearly ideology trumps science.

Chimpanzees are our closest living relatives, sharing nearly 99 percent of our DNA and a common ancestor six to seven million years ago. Chimps had by then already diverged from gorillas, yet today's chimpanzees and gorillas remain anatomically similar, adjusting for body size. Human skeletons, on the other hand, changed considerably since the divergence, making it likely that the common ancestor of chimps and humans likely looked like a chimp. (I thank Daniel Lieberman for this insight, though not everyone agrees [Almecija et al. 2021].)

Some descendants of that common root eventually became more humanlike than apelike, so we call them "hominins" and today assign most hominin fossils to one of two genera, either *Australopithecus*,

[1] The General Social Surveys, sda.berkeley.edu

whose species were bipedal with teeth like humans but retaining ape-size brains, or the later *Homo* genus, whose species have large brains like our own.

The earliest *Homo* – or human – fossils date to around the beginning of the Pleistocene Ice Age. One may as well think of the Age of Ice as the Age of Humans; they are concurrent.

Human Fossils

When Darwin published *On the Origin of Species* in 1859, there was no known fossil of a premodern human. The classic Neanderthal man, found in Germany's Neander Valley in 1856, comprising a skullcap with thick protruding brows along with two femurs (thighbones), was not then recognized as an extinct species (Jordan 1999). Hermann Schaaffhausen, who prepared the original description, thought the fossils to be remains of an ancient barbarous race of Europeans. Lacking an evolutionary framework, he had no notion that he might be looking at an earlier human species. One prominent pathologist argued that the skull belonged to a modern European who had suffered some deforming pathology like rickets, which thickened the brow. This became implausible after more Neanderthal fossils were found, but Neanderthals were still commonly viewed as a separate race, not a separate species. They did not look apish enough to be our primitive ancestor.

In retrospect, we see that fossils of Neanderthals were discovered earlier but not recognized for what they were. In 1848, for example, during work on fortifications at Gibraltar, workmen turned up a fine skull of elliptical shape, with a heavy brow ridge over each eye orbit and large nasal openings, which today would immediately be recognized as Neanderthal. But at that time there was no accepted context within which to interpret it as an evolutionarily ancient predecessor. Elsewhere, flint blades were found mixed with bones of mammoths and other extinct mammals, but again, their significance was not understood. By now there have been hundreds of Neanderthal fossils discovered in Europe and Southwest Asia, and there is a well-developed evolutionary framework within which to interpret them.

Darwin was not much bothered by the absence of fossil ancestors, believing that a sufficient method of study was to compare humans with living apes. (Similar logic is used today when inferring evolutionary relationships by comparing the DNA of living species.) Ernst Haeckel, Darwin's German disciple, thought fossils irrelevant for the further reason that ontogeny recapitulates phylogeny. To trace the evolution of humans, one need only watch the development of a human embryo through its "pre-human stages." Haeckel's views were enormously influential in the late nineteenth century. It was he, not Darwin, who drew phylogenetic trees to represent the evolution of species. In *The History of Creation* (1868), Haeckel showed a tree with 22 steps leading from the most primitive organisms to *Homo sapiens* as the top branch. The 21st step, immediately below humanity, was a postulated *Pithecanthropus alalus*, meaning ape-man without language. Presumed to have a brain about halfway in size between an ape and a modern person, this was Haeckel's "missing link" between them.

Dutch anatomist Eugene Dubois (1858–1940) decided as a young man to find fossils of the missing link. Traveling to Java, he searched until he discovered a half-million-year-old (by later dating) skullcap, two teeth, and a femur, which he called *Pithecanthropus erectus*, a modification of Haeckel's name to stress that this creature stood upright. Thus, stripped to its essentials, the discovery of Java Man sounds too fortuitous to believe, but it is true (Theunissen 1998; Swisher et al. 2000; Shipmen 2001).

Dubois's story is not quite so implausible if we add some detail. He reasoned that humans originated in the tropics because all extant apes live in the tropics. Lacking fur, humans are not well adapted for cold climates, and besides, no fossils more apelike than Neanderthals were turning up in Europe. Darwin had reached a similar conclusion and argued more specifically that Africa was the place of human origin because humans most resembled chimpanzees and gorillas, which live in Africa. Alfred Russel Wallace, whose views on natural selection were usually identical to Darwin's, disagreed. Believing humans were more closely related to orangutans and gibbons, Wallace thought their home in Southeast Asia the place of human origination.

It is likely that Dubois chose the East Indies, rather than Africa, for the practical reason that he needed financial support. The Dutch held colonies in the East Indies. Enlisting as an army surgeon, Dubois had the Netherlands government transport his family there and pay for living expenses while he sought fossils in his spare time. This sounds like the drunk who lost his keys and looked for them under the streetlamp because that was where there was light. Nonetheless, Dubois had cogent reasons for his choice. He saw the gibbon as the most generalized of living apes and therefore most resembling the human ancestor; gibbons live in the East Indies. Furthermore, the East Indies were known to yield fossils from the Pleistocene epoch, which Dubois regarded as an appropriate horizon for the presence of ancient human ancestors.

In 1887, Dubois arrived with the family at Sumatra and took charge of an army hospital. Four years earlier there had been an enormous volcanic eruption on the nearby island of Krakatoa, in the Sunda Strait between Java and Sumatra. It produced the loudest explosion in recorded history, destroying the island and 35,000 lives from the resulting tidal waves. Dubois did not say if that cataclysmic event influenced his perception of the location as a place of beginnings and ends.

Searching in his spare time, Dubois found mammal fossils including orangutans and gibbons. With this success he was able to secure government support including the help of fifty forced laborers and two military engineers. The subsequent method of digging, as in most archaeological expeditions, was to have a location selected by the person in charge while the heavy and tedious work of excavation and sifting was left to the workers and their supervisors. They searched while Dubois attended to his duties elsewhere. But after two years, the Sumatra results were disappointing because there was no difference between the fossils uncovered and the present fauna. In 1890, Dubois shifted his activity to Java, which seemed to have older fossils.

This was better. Workers found fossils of extinct species, and of extant species no longer in Java. In September 1891, during the second digging season, they found a primate molar in the fossil-rich bank of the Solo River at a stratigraphic level from the Pleistocene. The next month, they found at the same level an apelike skullcap, though clearly not an orangutan or gorilla. Two years later, they found a humanlike femur, again at the same level, roughly 15 meters from the location of the skull. Unfortunately, the fossil locations were not well marked, contributing to subsequent controversy about whether the skull and femur belonged to the same individual. Dubois was sure they did, declaring that his creature had a brain midway in size between an ape and a man, and a femur indicating that it walked upright like a human. Here was the missing link! It was an extraordinarily lucky find.

By providing an ancient and plausible ancestor for modern humans and Neanderthals and demonstrating that one could deliberately seek out fossil evidence of hominin descent, Dubois placed the study of human evolution on a completely new empirical foundation. Furthermore, his findings argued against Darwin's view that Africa was the "cradle of mankind." Dubois returned to the Netherlands in 1895 and defended to the end of his long life the claim that his fossils, and only his fossils, marked the midway point between humankind and its apish ancestor.

Homo erectus

Dubois's suggestion that humans originated in the East was widely adopted if usually retargeted toward the Asian mainland. Henry Fairfield Osborn (1857–1935), the imperious president of the American Museum of Natural History, saw Asia as "the chief theater of evolution both of animal and human life" (Mazur 2004). The Museum sponsored important expeditions to China and Mongolia, led by the dashing and publicity-seeking paleontologist Roy Chapman Andrews, often said to be the model for Indiana Jones. (Steven Spielberg denies it.) Andrews and his colleagues found excellent dinosaurs and, famously at the time, their eggs.

Despite his ballyhoo, Roy Chapman Andrews never found ancient human fossils in Asia. But J. Gunnar Andersson did. The story, as Andersson (1934) tells it, began in 1918 when a Chinese chemistry professor told him there was a promising fossil site at the town of Zhoukoudian, about 30 miles southwest of Beijing. Andersson visited the spot and verified the presence of small animal fossils but judging that the age of the deposits was not great, he turned to more promising diggings elsewhere.

In 1921, Andersson recruited a young Austrian paleontologist, Otto Zdansky, to come to China for a summer of excavations at Zhoukoudian, and again two summers later. Zdansky took the fossils he found to Sweden for cleaning and sorting. After considerable delay, he wrote Andersson from Sweden that he had two teeth from the 1923 season, which upon cleaning appeared to be human. (Teeth are the most common mammalian fossil.)

There was at that time a remarkable assemblage of Chinese, European, and North American scientists in Beijing, making it one of the leading centers of earth science in the world, a prominence little remembered in the West after the decimation of World War II and the Maoist takeover (Mazur 2004). Receiving photos of the teeth,

Andersson showed them to his colleagues in Beijing, who concurred that they were human teeth, apparently from the lower Pleistocene. American paleontologist Amadeus Grabau, then at Peking University, dubbed the find "Peking Man." In 1926, technical reports were published in the two most important scientific journals, *Nature* and *Science*, by Canadian Davidson Black (1884–1934), another member of the Beijing group, who identified the creature as "*Homo*?" The evidence was too uncertain to be more specific but fully adequate to bring in funds for more digging.

This was a chaotic time in China, with battling warlords and roving bandits in the area, but the diggers continued. More teeth were discovered, and in 1929 they hit the jackpot: a nearly complete skull! By the 1930s, several skulls had been uncovered.

Peking Man was the crown jewel of a decade of discoveries and a truly cooperative product of Chinese and Western workers, more important than almost any prior hominin find. The cave at Zhoukoudian provided the most complete and trustworthy picture of a *population* of proto-humans. Furthermore, features of lifestyle could be inferred from associated artifacts including stone tools, animal bones (presumably from prey), and carbon deposits suggesting the use of fire.

Franz Weidenreich (1873–1948), a German Jewish expert on Neanderthal fossils, moved to Beijing in 1935 to head the Peking Man project, a far better prospect than returning to Germany under Nazi rule. Excavations at Zhoukoudian were quickened to empty the site before it would be occupied by invading Japanese. When digging halted in 1937, pieces of at least 40 individuals had been uncovered, the largest collection ever made of a single population of extinct hominins.

Meanwhile, another German, Ralph von Koenigswald (1902–1982), found in Java several new skull fragments. Weidenreich and von Koenigswald met in Beijing to compare the Java and Peking fossils and concluded they were geological variants of a single species, to be called "*Homo erectus*." This unification was rejected by Dubois but has stood the test of time. *H. erectus* may be the most successful *Homo* species ever, its fossils found across Africa and from the Iberian Peninsula to China, and its longevity of two million years ten times that of *H. sapiens*.

The Chinese director of the survey, W. H. Wong, fearful the fossils would fall into Japanese hands, appealed to the US ambassador to have them safeguarded in America for the duration. After being photographed and castings made, they were packed into two large unpainted wooden crates and quickly transferred to an uncertain location, preparatory to shipment to the United States. Almost immediately, Japanese forces occupied the paleontology laboratory but did not notice the fossils were missing. (In 1942, when Japanese paleontologists sought them out, they found only casts.)

The crated fossils were to be evacuated with a marine detachment retreating from the American Embassy. On December 5, 1941, two days before the attack on Pearl Harbor, the marines boarded a special train to the coast to meet the American liner *S.S. President Harrison*, presumably taking along the crates. During those confusing days, the *Harrison* sank near the mouth of the Yangtze River, possibly destroyed by Japanese forces, possibly scuttled by her own crew to prevent capture. If the fossils were on board, they went down with the ship. However, it is doubtful that they were aboard because the *Harrison* seems to have sunk before its rendezvous with Peking Man.

One version of events has the Japanese taking the crated fossils while there were in transit and loading them on a barge that later capsized. In another version, the Japanese threw the bones away as worthless junk or sold them to Chinese traders as "dragon bones" used in traditional medicine. Other versions had them in the hands of a marine's widow, in Japan, in an American institution, or still in China. Some of these stories seem farfetched, but there is little hard evidence one way or another (Mazur 2004). Recalling the closing scene of the Indiana Jones movie, *Raiders of the Lost Ark*, I wonder if the fossils sit in some huge storehouse among thousands of other unmarked crates. In any case, they are lost except for their casts.

Out of Africa

In 1924, far away in South Africa, quarry workers blasting away at a limestone bed broke out the fossilized skull of a child, its milk teeth intact. The piece was turned over to Raymond Dart (1893–1988), professor of anatomy at the nearby University of Witwatersrand, who had no doubt it was a new kind of ape and coined the name *Australopithecus africanus*, or southern ape from Africa. One unusual feature was that the spinal cord appeared to leave the brain right beneath the skull, rather than toward the rear as in a quadruped ape, indicating that the child stood upright. Dart was convinced that

Australopithecus was a human ancestor. Unfortunately, the fossil was not well dated, which lessened its impact. Also, there was by this time a common belief that humans originated in Asia, not Africa, so many regarded the skull as an ape variant of no special importance.

One person who did take Dart and his fossil child seriously was Robert Broom (1866–1951), a retired Scottish physician who in 1934, at the age of 68, began searching South Africa for additional australopithecine fossils. By 1938, Broom had found enough adult specimens to swing opinion in Dart's favor. It was especially clear from adult teeth that the small-brained australopithecines had taken a step toward human dentition. The dating of these African man-apes showed convincingly that they were far older than *Homo erectus*. It was an easy presumption that some type of australopithecine was ancestral to some early *Homo*.

The most famous australopithecine today is Lucy, a 40 percent complete skeleton, 3.2-million-year-old, placed in the species *A. afarensis*. Discovered in 1974, her name comes from the Beatles song "Lucy in the Sky with Diamonds," frequently played during the fieldwork (Johanson and Wong 2009). Like all australopithecines, Lucy was found in Africa (Ethiopia) and had mixed human and ape characteristics. The facial proportions are apelike with a flat nose, protruding jaws, and no chin. Most significantly, Lucy and her kin had bipedal anatomy while still having ape-size brains, about one-third as large as a modern human brain. This infers that upright posture had evolved before large brains, a surprising conclusion in the 1970s when paleoanthropologists commonly supposed that a big brain preceded fully erect posture.

In 1978, Mary Leakey (1913–1996), matriarch of the famous Leakey family of paleontologists, and her team excavated at the Laetoli site in Tanzania a 24-meter line of hominin footprints in volcanic ash, cemented in place by soft rain and covered by a protective layer of more ash. The prints were dated to 3.7 million years ago, making them the earliest direct evidence of human bipedalism. They were produced by three individuals, possibly walking together, but in any case, demonstrating that hominins habitually strode upright. There was no sign of knuckle walking, as used by apes, nor do the feet have the mobile big toe of apes. Most experts attribute the footprints to an australopithecine species, *A. afarensis*.

By the 1980s, anthropologists had developed two competing views of the origin of *H. sapiens*. One was that our species had evolved in Africa, where the oldest hominin fossils were found, and then moved to other continents. This was dubbed the "Out of Africa" theory, after the romantic 1985 motion picture of that name starring Meryl Streep and Robert Redford. The other view, the "multiregional hypothesis," proposed that *Homo sapiens* had multiple origins, evolving independently in different parts of the world from local *Homo erectus* populations that had long before migrated from Africa. Thus, modern Chinese people evolved from *H. erectus* long resident in China, giving them a facially different look from modern people in the Middle East or in Africa, who had evolved from *H. erectus* already adapted to those regions.

The Picture Thus Far

It is tempting to think that we can eventually discover the exact chain of hominin transformations leading from the chimp-like root to what we are today, but probably we will never know which of many extinct species were our direct ancestors and which were evolutionary "dead ends." Several new hominin species have been named, sometimes from meager finds. Often these species are controversial, in part because paleoanthropologists who have invested years in the search exaggerate the uniqueness of their finds and their importance to human history. A common problem is that the range of variation within an extinct species is unknown. Consider that two specimens, unearthed by different anthropologists, may each be given a unique species name despite "really" being variants of a single species.

What can we say so far, especially when we shortly examine new DNA evidence? Africa is the cradle of humankind after all. The human line of succession from its chimp-like root to ourselves passed through an australopithecine stage and then through some variant(s) of *H. erectus*, eventually to modern *Homo sapiens* by 200–300 mya, all in Africa (Figure 9.1). This is an oversimplified picture, less a sequence of literal species transmutations than an approximation of changes occurring as some hominins evolved into us, their brains increasing in volume from about 400 cc to 1,200 cc; the skull becoming more globelike in shape; sloped foreheads becoming vertical to accommodate larger brains; faces flattening and chins jutting forward. One could envision the hominin head morphing in its entirely from an ape to a modern person. A famous hoax, Piltdown Man, created by a trickster who planted a human cranium near an ape jaw (stained to match), did not fit into this transition and was increasingly recognized as a fraud.

Figure 9.1 Skull facsimiles, from left: modern chimpanzee (*Pan troglodytes*), the australopithecine "Lucy" (*A. afarensis*); Peking Man (*Homo erectus*, China); and a modern human (*H. sapiens*)

The presence in East Asia (Java, China) of *H. erectus* fossils well over a million years old implies that their forbearers had earlier left Africa, with successive generations moving farther east. Water locked in glaciers and ice sheets left ocean levels much lower than today with wider expanses of coastal land. Groups of *H. erectus* may have taken a coastal route across the Arabian Peninsula, then through India, into the Bay of Bengal, occasionally rafting over shallow water. Their descendants could move north into China and south to Java. When modern *Homo sapiens* later migrated out of Africa by 50 thousand years ago (kya), but possibly in earlier waves too, they reached places already occupied by people, if different than themselves.

H. erectus in and out of Africa spun off more *Homo* species or subspecies, collectively referred to as *archaic humans*, now all extinct. Archaics are distinguishable from anatomically modern humans by having thicker skulls, prominent brow ridges, and lacking a prominent chin. The most famous of these, Neanderthal people (*Homo neanderthalensis*, or *Homo sapiens neanderthalensis*), may be descended from *H. erectus* through *Homo heidelbergensis* ("Heidelberg Man"). Neanderthal fossils are not known in Africa, so they likely appeared in the Ice Age conditions of Europe, perhaps accounting for their short and stocky build as a heat-saving adaptation to the cold climate. They hunted mammoths across Europe below the ice sheets, also living in Central Asia and the warmer Levant, where later they encountered *Homo sapiens*. Our kind may have had a hand in the Neanderthal extinction 40 kya.

Then there were the little people, an archaic species of dwarfed humans barely more than a meter tall, discovered in 2003 on the Indonesian island of Flores (*Homo floresiensis*, nicknamed "hobbits"). Nine individuals have been found as well as stone tools, with dates from 60–100 kya.

The hobbits preyed on dwarfed elephants that lived on the island. The status of *H. floresiensis* as a species, and how it relates to other hominins, is not resolved (Argue and Groves 2017).

Perhaps the most interesting of the recently discovered archaic species are the Denisovans (tentatively, *Homo denisova*). In 2010, scientists announced the discovery of an undated finger bone of a juvenile female found in the Denisova Cave in Siberia, which had also been inhabited by Neanderthals and modern humans. Mitochondrial DNA retrieved from the finger showed it to be genetically distinct. Shortly afterward but far away, a jawbone with some molars was discovered in a cave at high altitude on the Tibetan Plateau. Protein analysis identified it as another Denisovan specimen (Gibbons 2019). These two finds showed little of the Denisovan skeleton but demonstrated that its geographical range and altitude tolerance were considerable. Current dating has Denisovans and Neanderthals splitting from the modern human lineage one million to 800 kya, and from each other 470 kya to 380 kya (Reich 2018). A recently analyzed genome extracted from a hominin found in Russia shows the individual had a Neanderthal mother and a Denisovan father (with some Neanderthal ancestry) while giving no hint about joint childrearing. The hybrid find, if properly identified, was extremely lucky or suggests that interbreeding was not a rare occurrence (Slao et al. 2018).

Other archaic human species are known (e.g., *H. rhodesiensis*, *H. antecessor*, *H. naledi*, *H. luzonenesis*), and we may expect more to be discovered. Their relationships to one another are likely to remain obscure, so it is uncertain how to place them in the larger scheme of things. My guess is that some will be lumped in with *Homo erectus*.

Hominins did not evolve in a smoothly direct line, as often portrayed in evolutionary trees of decent. The "tree" is more like an unruly bush with branches going diverse ways, some interbreeding, all but one now extinct. There were at least five *Homo* species (or subspecies) alive during the late Ice Age, as recently as 100 kya: modern *Homo sapiens* (us); Neanderthals; Denisovans; *floresiensis* "hobbits," and the last of the long surviving *Homo erectus* (Rizal et al. 2019). There may have been others, still unrecognized.

DNA Revolution

Just as carbon dating revolutionized paleontology, so again did the analysis of DNA from living and fossil people. The genome, the genetic information that everyone inherits from its two parents, is chemically encoded in long strands of the molecule DNA. Chromosomes are threadlike stretches of DNA found in the nucleus of most living cells, most of them combining sequences from both parents. Humans normally have 46 chromosomes arranged as 23 pairs.

We speak of a "human genome" as if every human had the same genes, which is not true except for identical twins. The genome of every non-twin is a unique combination of their parents' genes and can serve as an individual ID tag. But the genetic variation from one person to another is small, compared to the 20,000-or-more genes estimated to be common across humans, so any person's genome may be regarded as typical of its species.

Francis Crick, James Watson, Rosalind Franklin, and Maurice Wilkins showed in 1953 that the genome is encoded in twin chains of DNA, spiraling around one another. Each DNA strand (or chromosome) within the "double helix" is a long chain connecting in variable order four kinds of nucleotides. Think of these nucleotides as letters denoted A (adenine), C (cytosine), G (guanine), and T (thymine). Short stretches of a chromosome, typically about 1,000 letters long, are what we know as genes. (Much of the chromosome is not genes at all but repetitive sequences of letters that have no discernible function and are called "junk DNA.")

Beginning in the 1990s, there was a very elaborate and costly international collaboration, the Human Genome Project, to sequence the entire human genome of billions of letters.

Subsequently the process has been simplified, automated, and made vastly less expensive, so that anyone can now get their genome sequenced from a commercial service, learning their own line of descent, and finding living relatives otherwise unknown (or occasionally learning one is not genetically related to someone thought be a blood relative).

Geneticists used to think of genes as if they were beads strung along a chromosomal string. Today, we know genes as short sections of a chromosome, a particular sequence of nucleotides, from which the cells construct a particular protein. The technology for reading nucleotides has progressed so rapidly that today entire genomes are sequenced cheaply and quickly, even from fossils if enough intact DNA can be recovered.

Slight changes in a genome may occur through mutation, perhaps because of an error in the cell's copying mechanism or by hits from cosmic rays. These are random events. If a population were to be split in half and separated for a long time, mutations subsequently occurring in one half would differ from those occurring in the other half. The mutations in these two halves might accumulate sufficiently to be detected by a gene sequencer. Assuming a constant rate of random mutation, one could estimate how long ago the groups split.

A small portion of the genome is contained in mitochondria, tiny organelles in the cell outside its nucleus. Mitochondrial DNA is passed down from a mother to her children. (A man's mitochondrial DNA is inherited from his mother but not passed to his children.) Aligning the letters of a person's mitochondrial DNA with that of their great grandmother would show the same sequence of letters except for an occasional mutation.

In one of the earliest applications of DNA sequencing, Allan Wilson and his colleagues compared hundreds of letters of mitochondrial DNA from people around the world and were able to construct a mitochondrial family tree from numbers of mutations. They found that the human branch that left the main trunk earliest contained people found today only in sub-Saharan Africa. All non-Africans descend from a later branch of the bush. Based on the rate at which mutations accumulate, Wilson and his followers estimated that the most recent African female ancestor of all of us, dubbed "Mitochondrial Eve," lived 200,000 years ago. This strongly favors the Out of Africa theory. If today's humans had developed independently from diverse groups of *H. erectus* living in various parts of the world, as proposed by the multiregional hypothesis, then the most recent common mitochondrial grandmother would have been an *H. erectus* woman who lived two million years ago.

Naming this female ancestor after the biblical matriarch does not mean she was the only living female at the time. It simply means that the mitochondrial DNA of her female neighbors eventually disappeared for lack of a surviving daughter. Still, these neighbors might have passed on non-mitochondrial DNA through their children.

There are far more genes in the cell's nucleus than in its mitochondria. The nucleus has 23 pairs of long chromosomes, mixing genetic contributions from both parents. Twenty-two of the chromosome pairs have no sexual function. One pair does, determining the sex of the offspring. If that pair has two X chromosomes, the newborn is a female. If the pair has one X chromosome and one Y chromosome, it becomes a male. Since females lack a Y chromosome, the sex of the child is determined by the father, who may pass down either an X, hence a daughter, or a Y, hence a son.

Just as we can trace back to the most recent woman who passed down all of today's mitochondrial DNA, there is an analogous most recent man from whom all of today's Y chromosomes came. Only males have a Y chromosome, so only males pass them to their sons. It was inevitable, I suppose, that this ancestor would be called "Y-chromosome Adam." Recent studies indicate he lived 100,000 to 200,000 year ago (Callaway 2013).

Unfortunately, biblical names suggest Adam and Eve knew each other and mated, but that is not a correct inference. They probably lived far apart in time and space. It is simply that for one reason or another, perhaps by chance, they were the most recent humans whose mitochondrial genes or Y chromosome was passed down to the rest of us. Their contributions are only a small part of the human genome. Most of it came from other people. Adam and Eve were not the progenitors of *Homo sapiens*. Our species existed before they were born.

These early results, based on genetic differences among living people, were impressive, catching the public interest, though results had limited scientific impact. The importance of DNA has since ramped up by sequencing (inexpensively) the whole genome rather than limited pieces, and especially doing it for archaic humans. Svante Pääbo (1955–), a Swedish geneticist now at the Max Planck Institute for Evolutionary Anthropology in Leipzig, Germany, pioneered many of these techniques in his long project to sequence the Neanderthal genome, an incredibly difficult but successful endeavor (Pääbo 2015). Comparing human and Neanderthal genomes, Pääbo produced an evolving list of about one hundred thousand places where nearly all of today's humans carry genetic changes that are absent in Neanderthals. Presently it seems unlikely that gene changes will explain how modern human behavior evolved. The greatest short-term benefit of the ancient genome revolution is historical, i.e., how our lineage changed from *Homo erectus* to modern

Homo sapiens, and furthermore how different populations of modern *H. sapiens* migrated and mixed with one another.

Specifying the family tree (or bush) of the various *Homo* species will improve as new fossils are found and techniques improve for extracting analyzable amounts of DNA. (DNA deteriorates, so very old fossils may have none to be extracted.) It presently appears that Neanderthals, Denisovans, and *sapiens* derived in parallel lineages from *Homo erectus*, possibly through an intermediate European species, *H. heidelbergensis*, though a more definitive genealogy must await new data. In any case, Neanderthal remains are dated in Europe as early as 400 kya.

According to David Reich (2018), a leading researcher of ancient DNA, studies so far show that the main ancestral population of modern humans, Neanderthals, and Denisovans separated from the superarchaic *Homo* lineage some 1,400 to 900 kya. Denisovans departed the lineage of modern humans one million to 800 kya. Neanderthals separated from the line of modern humans between 770 and 550 kya. Neanderthal-Denisovan populations were genetically split by 470 to 380 kya. The oldest fossils looking like anatomically modern humans, found in Morocco, are dated to about 300 kya.

Whole genome analysis shows that modern humans, Neanderthals, and Denisovans sometimes interbred, producing fertile offspring. Indeed, one of the biggest lessons of the young field of ancient genome analysis is that there was a lot of interbreeding among the various hominin "species." This may contradict the criterion that distinct species cannot mate to produce fertile offspring, though the issue can be finessed as we have always done, being long aware of living "species" that do occasionally produce fertile hybrids.

In times past, different hominid species sometimes lived in proximity, perhaps hunting one another, or mating, or both. *Homo sapiens* certainly had sex with Neanderthals because those of us who are not African typically carry about 2 percent of Neanderthal genes. The gene-testing company 23andMe reports that I have 240 Neanderthal genetic "variants" out of 2,871 variants tested. (Neanderthals did not live in Africa so their genes are not found in Sub-Saharan people, excepting small amounts which may have come from non-Africans re-entering Africa.)

Each carrier of Neanderthal genes does not carry the same ones. Collectively, a considerable portion of the full Neanderthal genome still exists in today's human population. We interbred with Denisovans too, especially among Oceanians (Bergstrom et al. 2020). We have no idea if this sexual interaction was amiable or agonistic.

Human Migration

The picture of human migration remains blurry but is becoming clearer. From fossil and DNA evidence, we know that hominins evolved in Africa and eventually migrated out of that continent. Some who remained evolved into modern *Homo sapiens* by 200–300 kya. By that time, *Homo erectus* had been living in Java and China for over a million years, so their progenitors, some early form of *Homo*, must have left Africa much earlier, perhaps two million years ago (Reich 2018).

Neanderthals and Denisovans evolved in Eurasia by 400 kya, so their ancestors must have left Africa before the evolution of *Homo sapiens*. When modern *Homo sapiens* finally migrated out of African, probably in several waves but certainly by 50 kya, they encountered in the Levant and elsewhere people not quite like themselves but close enough for mating. (The new arrivals were once called "Cro-Magnon" but are now known simply as Europe's early modern humans.) For several thousand years, *Homo sapiens* and Neanderthals lived in propinquity, until the Neanderthals went extinct by 30 to 40,000 years ago.

As glaciers grew, removing water from the ocean basins, many islands and continents were temporarily connected by land (Figure 9.2). *Sapiens* in Southeast Asia were able to walk or make short sea crossings all the way down the Malay Peninsula to Sumatra, Java, Bali, and Timor. From Timor, they would see open ocean, not knowing that Australia was over the horizon. Nonetheless, they did occupy Australia, apparently by 65 kya, requiring a deep ocean crossing of about 90 km to reach Australia (Clarkson et al. 2017). Millennia later, their descendants and people from Southeast Asia made long-distance ocean voyages to reach far flung islands of the Pacific, a feat that still seems incredible.

Up north, modern *Homo sapiens* reached Siberia around 20 kya, spreading across northeastern Asia. The Western Hemisphere was last to be occupied because passage through Beringia was blocked with ice and snow. As the climate warmed, *Homo sapiens* could reach North America, but once in Alaska were confronted by an unbroken wall of ice blocking movement farther south across Canada. Only when the ice

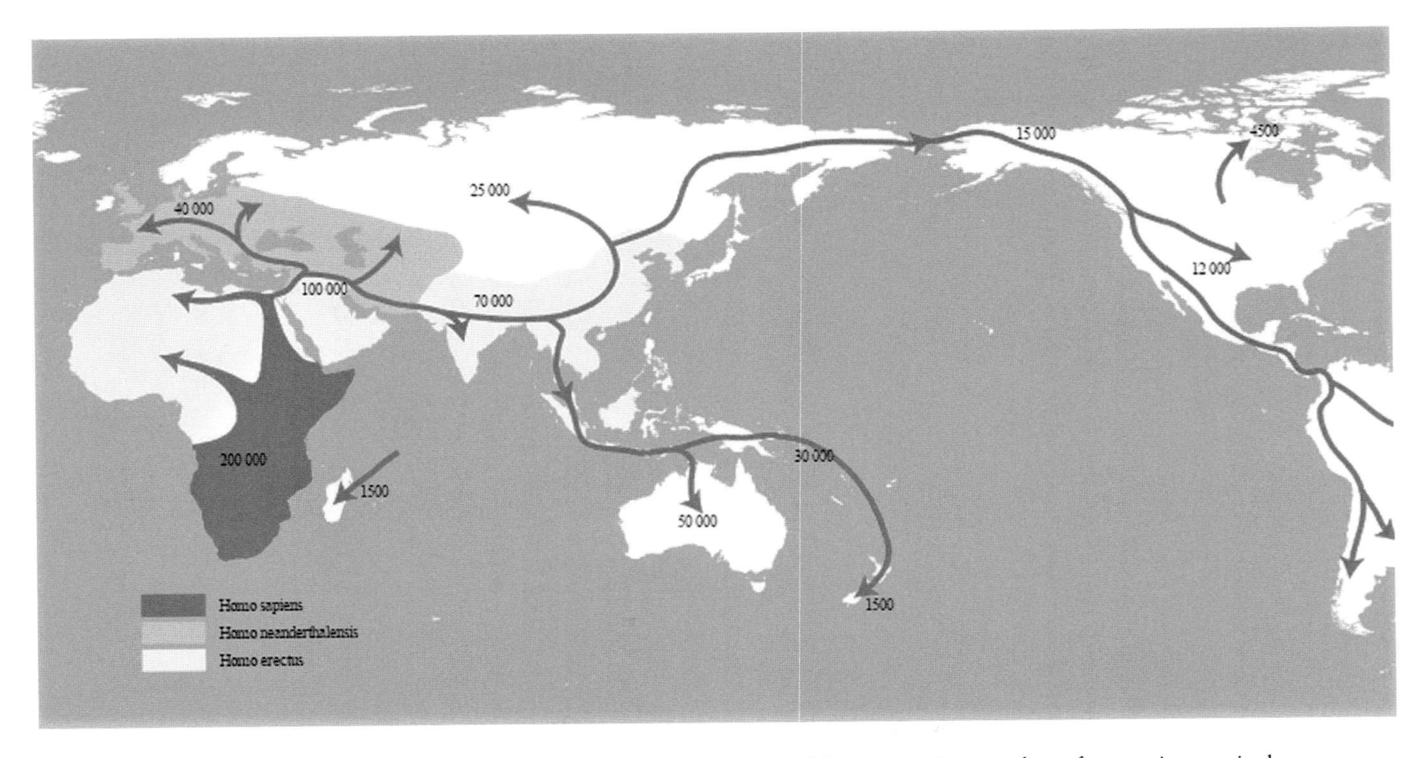

Figure 9.2 Proposed routes of migration of *Homo sapiens* out of Africa, with conservative number of years since arrival.
Source: Wikipedia Commons

began to melt back, about 15,000 years ago, did the bottled-up Alaskans have access to the south. An ice-free corridor opened between the continental ice sheet in eastern Canada and the glaciers that mantled the western mountains. Through this ice-chilled corridor, or along a coastal route, moved a wave of human migration, within 2,000 years reaching the southern tip of South America (Raymo and Raymo 2007).

10 HOW DID EXTINCT HOMININS BEHAVE?

In many respects the living members of the primate order seem to form a natural ladder, from primitive to more advanced, or specialized, types. This remarkable array provides us with living species that preserve many features that characterized primates of earlier epochs. This diversity of form can give us some idea of the pathways of primate evolution.

John Fleagle (1999: 81)

Soft parts of animals rarely fossilize. Knowing the DNA of an extinct animal would in principle give us a blueprint of its organs and behavior but in actuality tell little about the creature's physiology and actions. How, then, can we reconstruct soft parts and behaviors of species from the distant past? There is an easy solution in a few cases because some ancient species are still with us, more or less.

Walking the Delaware shore, I occasionally find a dead or barely living horseshoe crab (*Limulus*), the classic "living fossil," scarcely different from specimens dated 300 million years ago (mya). Today's scorpions also look nearly the same as ancient fossils. Living fossils are similar but not identical to ancient fossils and are therefore not truly ancestors of other living species. Indeed, orthodox Darwinians insist no species alive today evolved from another living species. A common error of this kind is to say humans evolved from chimpanzees, or, from the chimp's point of view, that chimpanzees evolved from humans. Neither claim is correct. The lineage of today's chimpanzee is exactly as old as our human lineage. The correct statement is that humans and chimps have a common ancestor, now extinct.

If one is willing to deviate a bit from orthodoxy, and nearly everyone is, then living fossils become valuable for reconstructing life's history. Since hard parts of the modern horseshoe crab look like ancestral fossils, it seems likely that their soft parts looked similar too. On these grounds, hardly anyone objects to using the living animal as a model for its extinct ancestor, tracing its nervous system, digestion, reproduction, and so on. Furthermore, living horseshoe crabs can be watched, their habits studied for indications of ancestral behavior. They eat a wide variety of food, catching worms and mollusks and chopping them into small bits with the bases of their legs, also scavenging bottom debris. They tolerate large fluctuations in salinity and are often the last species driven from an estuary by human pollution. Leaving the water to lay their eggs in nests on the beach, they can survive for days in the air and sunshine. Altogether they are hardy generalists, able to live in diverse conditions, and probably their ancestors were too, which may explain why they survived with little modification (Eldredge 1991). Species that look alike do not necessarily act alike, but to cautiously assume they do is a reasonable starting point.

Most mammals evolved rapidly, leaving few species alive today that closely resemble ancient forms. Therefore, we must go farther out on a limb, modeling the behavior of long-gone mammals after that of somewhat dissimilar living relatives. We may claim, for example, that since all living mammals nurse their young, all have strong mother-infant bonds, and all their juveniles play, then probably these behaviors were present in the most recent common ancestor of living mammals. This assumption – in essence that behavior and soft parts are conservative within lineages – is common in theorizing about extinct animals.

What Is a Primate?

The zoological definition of the Primate order is complex, but the most important thing we have in common is that we are tree-dwellers. or are descended from them, with features useful for that habitat. Five digits at the end of each limb, including opposable thumbs are ideal for grasping branches. Our digits end in nerve-filled pads, at least some of them backed by flat nails rather than claws, making them effective sense organs. Primates have sharp eyes, in preference to keen noses, and our eyes are set inside protective bony orbits. Rather than looking sideways, like a dog, primate eyes look forward, producing overlapping fields of view that the brain combines into depth perception, useful in moving through the three dimensions of a forest canopy.

Brains are larger in primates than in other mammals of similar size, and more elaborate in design. Usually, pregnancies produce one offspring at a time, the mother carrying it for months before giving birth. Babies grow slowly and are highly dependent on maternal care – features closely connected to the development of complex social organization. The main task of primate mothers is to protect their vulnerable young and socialize them into primate society. This requires an unusually large investment of time, energy, and emotion, but it has the advantage of enabling each new generation to learn from the prior generation.

A Series of Living Primates

Living primates fall into six "natural groupings," reflected one way or another in all primate taxonomies (Martin 1990):

1. Lemurs (Madagascar; 18 genera, 35 species)
2. Lorises (Africa, South and Southeast Asia; 5 genera; 11 species)
3. Tarsiers (Southeast Asia; one genus, 4 species)
4. New World monkeys (South and Central America; 16 genera, 64 species)
5. Old World monkeys (Africa, South and Southeast Asia; 16 genera, 73 species)
6. Apes and humans (Africa, South and Southeast Asia, not counting recent migrations; 5 genera, 12 species)

The first three groups – lemurs, lorises, and tarsiers – are traditionally called "prosimians," as distinct from simians (i.e., monkeys, apes, and humans). See Figure 10.1 for exemplar species.

Taxonomic definitions aside, to the average viewer most monkeys, whether from the Old World or New World, are visually recognized as monkeys. The great apes, larger and more humanlike, have a considerably different appearance. Prosimians often seem more like squirrels or raccoons than monkeys, and many zoo-goers do not recognize them as primates. On virtually all dimensions, the prosimians

Figure 10.1 Representative primates, clockwise from upper left: prosimian (lemur, by Ian Kelsall); New World monkey (squirrel monkey, *Saimiri sciureus*, by Miguel Collado); African ape (chimpanzee, *Pan troglodytes*, by Karen Lau); Old World monkey (macaque, by 2Photo Pots). Photos provided by Unsplash

are least like us, the apes are most like us, and monkeys hold an intermediate position.

There is considerable agreement about the broad outline of primate phylogeny, diagrammed in Figure 10.2 (Fleagle 1999). The earliest recognizable primates were prosimians, not exactly like extant prosimians but in many ways like today's lemurs and lorises.

Fossils indicate the existence of apes, derived from some kind of Old World monkey, by 20 mya. Relatively quickly, the ancestors of gibbons and siamangs split off the "main line" to great apes and humans. Somewhat later, the first orangutans diverged from the progenitors of today's African apes and humans.

We cannot go so far as to line up today's primate species in an unambiguous developmental path, because every surviving lineage adapted uniquely to its peculiar ecological history. But while natural selection produced modifications that improve genetic fitness, it is otherwise conservative (Blomberg and Garlan 2002). Amidst the diverse adaptations of extant taxa, there are central tendencies reflecting the preserved heritage of common ancestors. By tracing these modal features from prosimians, though New and Old World monkeys, to

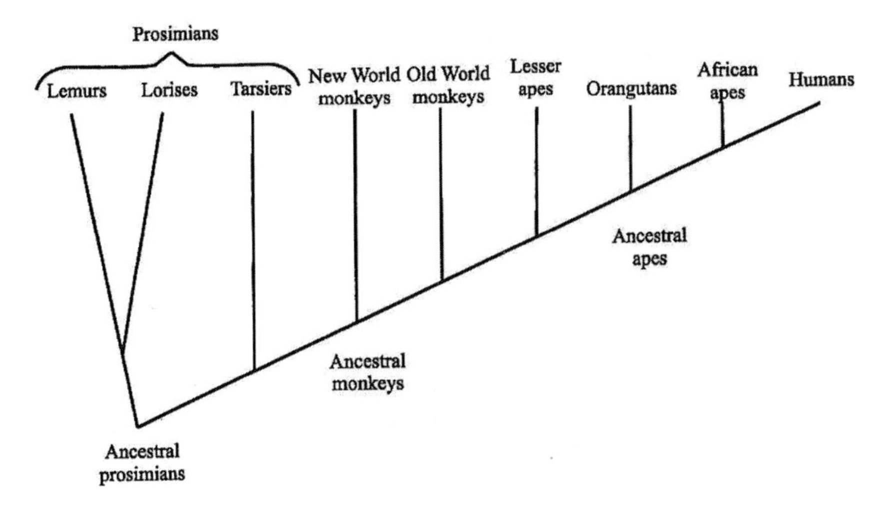

Figure 10.2 Phylogeny of today's major primate groups

African apes, to humans, we discern the approximate evolution of prehistoric human qualities.

The degree to which this primate series shows the true course of human evolution depends on the amount of change in each lineage since it diverged from the main line of human descent. If prosimians evolved slowly, then living prosimians are good representatives of their ancestors. The picture is complicated because lineages change at different rates. Judging from comparative morphology, apes diverged much faster and further than prosimians from their last common ancestor, but the African apes have not changed much since their split from hominins. Considering body size differences among gorillas, chimpanzees, and bonobos, they are all anatomically similar. Fleagle likens gorillas to overgrown chimpanzees, and bonobos to juvenile chimps (1999: 253). In the last chapter, we inferred that the common ancestor of chimps and humans was anatomically like a chimpanzee. The earliest hominin fossils all appear chimp-like. On these grounds, living chimpanzees seem a suitable model for the common chimp-human ancestor.

R. D. Martin, who is especially critical of incautiously treating extant species as "frozen fossils," nonetheless justifies the interpretation of major groups of living primates as being at different "grades of evolutionary development" (1990). The visual continuity in physical appearance, evident in adults (Figure 10.3), is even stronger when

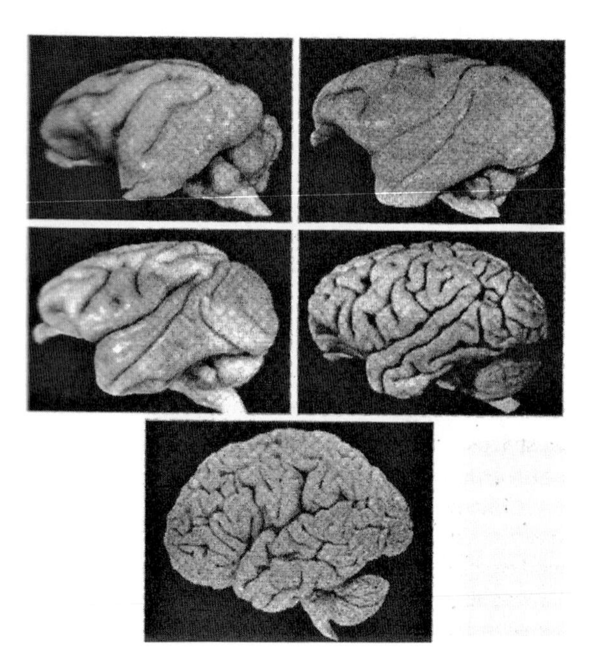

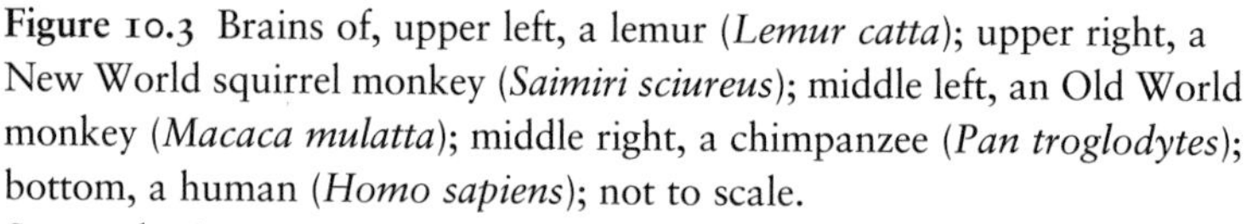

Figure 10.3 Brains of, upper left, a lemur (*Lemur catta*); upper right, a New World squirrel monkey (*Saimiri sciureus*); middle left, an Old World monkey (*Macaca mulatta*); middle right, a chimpanzee (*Pan troglodytes*); bottom, a human (*Homo sapiens*); not to scale.
Source: brainmuseum.org, supported by the National Science Foundation

juveniles are compared. Body size generally increases across the primate series, and the brain enlarges relative to the size of the body.

Arranging living primates in a quasi-evolutionary series allows us to make educated guesses about the soft parts and behavior of extinct ancestors. For example, what did ancestral brains look like? It is possible to make an endocast from the interior of a fossil skull, showing the overall shape of the brain that once fit into it, and even some features of its surface, but this is barely a glimpse. The comparative anatomist might take a different approach, using the evolutionary relationship of living primates, whose brains can be directly observed (Figure 10.3). It would not be audacious to guess that the brain of an extinct hominin looked intermediate between the chimpanzee and human brains pictured.

There is a general prolongation in the stages of life across the primate series. The modal durations of gestation, infancy (until eruption of milk teeth), the juvenile period (until eruption of permanent teeth, and attainment of sexual maturity all increase, though with considerable variation within the major groupings. This lengthening of dependency and postponement of sexual independence is accompanied by more extensive and flexible socialization of the young.

Numerous aspects of social living appear along the primate series and are conserved. Intelligence, measured in several ways, becomes more humanlike as we advance through the primate series (Mithen 1996). Apes are more adept than monkeys at learning artificial sign "languages" and symbol systems invented by humans. Chimpanzees are like humans in several aspects of "social intelligence." They look into one another's eyes in an affectionate context, practice deception, allow others to take food, probably share food, and cooperate in various ways including hunting. Monkeys and prosimians do not show these behaviors (Matsuzawa 2001).

Monkey-level intellect is sufficient for intergenerational passage of elementary cultural information. In one famous example, primatologists provisioned a troop of Japanese macaques (*Macaca fuscata*) living by the ocean shore with sweet potatoes and wheat. The monkeys invented customs for eating these foods. The macaques soon learned to carry handfuls of mixed sand and wheat into the sea. The sand sank, making it easy to eat the separated wheat. They "washed" sweet potatoes in seawater, perhaps imparting a salty flavor. These customs were still practiced years later (Hirata, Watanabe and Kawai 2001).

Young chimpanzees, watching their mothers, learn to use tools, including prying with sticks, throwing stones, and wiping themselves with leaves. Furthermore, they learn to *make* tools out of natural objects, such as chewing leaves to increase their effectiveness as sponges for sopping up drinking water (Takeshita and van Hooff 2001). They strip and trim twigs for use in catching termites, a favorite food. The chimp inserts the modified twig into a crevice containing the insects; the termites walk up the twig, and the chimp licks them off. These examples may not impress an industrial machinist, but they were discovered in the postwar decades while many scholars considered the ability to use tools a defining characteristic of being human. When a young Jane Goodall reported to her mentor, Louis Leakey, that she saw chimps modifying natural objects for utilitarian tasks, his now iconic response was, "Now we must redefine tool, redefine Man, or accept chimpanzees as human" (Goodall 1986).

Tools

Australopithecine sites are barren of stone tools or signs of tool use. One exceptional finding (to date) is cut marks on bovine bones dated 3.4 mya, so perhaps *A. afarensis* used stone cutters and ate meat (McPherron et al. 2010). But this interpretation is controversial. In contrast, paleoanthropologists commonly find stone tools at sites occupied by ancient *Homo* species. Tools or fragments of tools may litter the ground, being far more plentiful than *Homo* teeth or bones. This led to the general impression that tool use was a distinctive characteristic of the genus *Homo*, an important change from australopithecine behavior (Shea 2017; O'Brien, Buchanan, and Eren 2018).

Some years earlier, in 1964, the British-Kenyan anthropologist Louis Leakey (1903–1972) and his colleagues described hominid fossils from the Great Rift Valley in Tanzania, proposing they were a new species of our own genus (Leakey, Tobias, and Napier 1964). They chose its name, *Homo habilis*, or "handy man," presuming it was the maker of primitive stone chopping or cutting tools discovered at nearby Olduvai Gorge (Figure 10.4). They thought it was the first hominin to use (and make) tools. *Habilis* did not have a large brain (ca. 600 cc), but its namers believed it deserved inclusion in *Homo* because of its dexterity at toolmaking.

A few more *habilis* specimens were found, but the status of the species remains controversial. Some experts doubt it is human, classifying *habilis* as an australopithecine because of its small brain. Some think it should be a genus of its own, transitional from *Australopithecus* to

Figure 10.4 Oldowan chopper.
Source: Wikipedia, released into the public domain by its author, Locutus Borg

Homo, or perhaps it is simply an early variant of *Homo erectus*. Whatever it should be called, did tool use and manufacture begin roughly with the arrival of *Homo*, as Leakey and his colleagues thought? Fossils cannot provide a clear answer because if earlier tools were made of bone or wood, they would have disintegrated, leaving nothing to find.

A different approach turns to the primate series, where ancient hominins are bracketed between chimps and modern *Homo sapiens*. Knowing that chimps use and make tools, it would be extraordinary if early hominins did not. From the dearth of stone tool remains, it is doubtful that australopithecines made much if any use of them, but very probably they did modify organic materials for practical purposes, at least as well as chimps do.

Paleontology is on clearer ground with *Homo* species, which not only made stone tools aplenty, but followed different cultural traditions. The earliest tradition (or "industry"), first identified at Olduvai Gorge, is called "Oldowan" after the type site. Its tools are crude, barely recognizable as manufactured implements (Figure 10.4). They are either blocks of stone from which flakes have been removed, or the flakes themselves. They appear from roughly 2.6 mya until 1.7 mya, their presence spreading from Africa to South Asia, the Middle East, and Europe.

A newer industry, more varied and with finer workmanship, essentially replaced Oldowan technology in Africa, then moving across southern Eurasia. Called "Acheulean" after a French type site, it is closely associated with *Homo erectus*. Obviously, the use for each tool cannot be known with certainty, so we have only reasonable guesses. The kit apparently includes choppers, scrapers, awls, hammerstones, and a characteristic two-faced hand axe that is its hallmark (Figure 10.5).

It is tempting to assume that Oldowan tools were the acme of *H. habilis* intellectual ability, that *Homo erectus* evolved a higher mental capacity sufficient to develop Acheulean tools, but that was its limit because the kit remained relatively static for 1.5 million years. A still finer tool industry, the Mousterian, is associated with the later appearance of Neanderthal people, possibly reflecting their better wired brains (Figure 10.6). In effect, this assumes that each new species quickly reached the limit of its intellectual capability and went little further, excepting minor regional variation. In this view, progress, at least in tools, requires evolution to a higher mental ability, which only then allows a better tool kit to take over. As various archaic *Homo sapiens* arrived, there was a proliferation of toolmaking techniques.

Figure 10.5 Acheulean handaxe from Kent, England, worked symmetrically and on both sides, apparently by *Homo erectus*. *Source*: Wikipedia Commons

This view cannot be entirely correct. Brain power works within the constraints and opportunities of the culture within which it is operating. Isaac Newton (1642–1726), possessing one of the finest minds in Europe, calculated the year of Creation just as Archbishop James Ussher had, reaching a similar answer, which to modern scientists is absurd. (A believer in astrology, he also cast horoscopes.) Newton's failure was certainly not due to his brain but to the culture within which he (and Ussher) lived, which took for granted that the Bible is authentic and true.

During the two-plus million years that *Homo* evolved, not only mental capacity but cultural capacity has increased, surely helped by the acquisition of language that transmits and accumulates knowledge and ideas from one generation to the next. Our brains today figure out how to build computers and reach the moon, but an equally good or better brain in Newton's time would not have even conceived of such technologies. Same brains, different cultures. The industrialization of much

Figure 10.6 A thin, sharp Mousterian scrapper from France, flint, attributed to Neanderthals, 300 to 40 kya.
Source: Wikipedia, licensed by Didier Descouens as part of Project Phoebus under Creative Commons attribution

of the world's population over the past three centuries cannot be identified with any physical improvement in the brain, but is explained by changes in the cultural landscape within which brains were thinking (Diamond 1997). Stasis of prehistoric tool traditions may as well be attributed to static cultures as to genetic limits of the brain's wiring.

11 LIFE IN THE PALEOLITHIC

Sir John Lubbock (1834–1913) was an English banker, baronet, politician, and amateur archeologist. He was a young neighbor and friend of Charles Darwin, a pallbearer at the great man's funeral. In 1865 he published *Prehistoric Times, as Illustrated by Ancient Remains*, perhaps the most influential archeological textbook of the nineteenth century (Mithen 2006). In it, he coined the terms "Paleolithic" and "Neolithic," or Old Stone and New Stone ages. His emphasis on stone was practical. Stone implements are the least destructible and most easily found relics of ancient peoples. Certainly, he was interested in the full cultures of prehistory, not simply their stony tools, but stones provided the best access, at least until humans leaned how to smelt metals.

Lubbock divided prehistory into four epochs – the Paleolithic came first, when humans lived alongside mammoths, cave bears, and other extinct animals. Its stone tools, found with extinct animal bones, were crude, sometimes barely recognizable as tools.

His next epoch was the Neolithic, a time of "beautiful" weapons and instruments made of flint and other kinds of polished stone, when the only metal found was gold, sometimes used for ornaments. (Today's usage of "Neolithic" is different, referring less to the quality of stone implements than to the beginning of agriculture.) Third came the Bronze Age, after humans learned to smelt copper and tin and to mix copper with tin (or arsenic) into the alloy bronze, which held a sharp edge for arms and cutting instruments. Fourth was the Iron Age, when that harder-to-smelt and far stronger metal supplanted bronze for arms, axes, and blades.

Writing after savants had "burst the limits of time," Lubbock knew the Paleolithic must have lasted far longer than the 6,000 years allowed by the Bible. Modern dates for the beginning and end of the Paleolithic, from the first appearance of stone tools to the beginning of metalworking and agriculture, are about 3,000,000 years and ten thousand years ago, or 99 percent of the time the genus *Homo* has existed, virtually concurrent with the Pleistocene Ice Age. In North Africa and Eurasia, the beginning of the Bronze Age is today dated between 8,000 and 5,000 years ago.

As the Paleolithic epoch was popularized among European archeologists, Americans wondered if there had been a Stone Age on their own continent too, entailing the question of how long humans were present there. The geological strata of America are similar to Europe, implying similar chronology, so why not similar antiquity of humans? Other archeologists adamantly claimed that people did not arrive in North American earlier than a few thousand years ago. This devolved into acrimonious controversy, fierce even by academic standards. Discoveries in America of human bones in proximity to mammoths and other extinct animals were refuted as evidence for an Ice-Age presence, arguing that close proximity could be accidental and did not imply the humans and animals lived at the same time.

This bitter dispute came to conclusive resolution at Folsom, New Mexico in 1927 after archeologist Jesse Figgins found a flint projectile point buried between two ribs of an extinct type of bison. Rather than extracting the point, Figgins removed the assemblage intact so any sceptic could see their association (Meltzer 2015). In the following years, it became broadly accepted that prehistoric Americans did have a stone age during the late Pleistocene. However, it was also generally accepted that humans were not present in America for a very long time, so the American Stone Age was not nearly as long as that of the Old World.

Lubbock never intended his epochs as universal stages. In sub-Saharan Africa, the Paleolithic was followed directly by the Iron Age. In South America, the Stone Age continued until 4,000 years ago when gold, copper and silver were smelted, but there was no widespread use of bronze or iron, although the technology existed (Easby 1965). During the metal ages, stone tools never wholly disappeared, perhaps because they were cheaper than bronze or iron.

I envision a Martian spaceship visiting Earth every 10,000 years, with anthropologists on board conducting a surveillance from orbit. For most of *Homo*'s existence, they would have reported back that human life is pretty much the same as it was on their last visit. People still live in small nomadic groups, sustaining themselves off the land by hunting, gathering, and scavenging. Occasionally they meet other groups, perhaps relatives, camping together for a few days of trading, partying, and exchanging mates, sharing food around a fire, telling stories of times past, giving information about places where prey animals or flint are abundant. Perhaps they would pray together or celebrate some ancient rite. Then each group would depart, resuming its independent daily existence.

This was life in the Paleolithic, pretty much the same from generation to generation, as we might see in periodic visits to chimpanzee habitats. Of course, there would be environmental perturbations, climatic changes, perhaps strife with a neighboring group, but overall a barely changing lifestyle, the kind of stasis seen in the long Oldowan and Achulean tool traditions. To a first approximation, the Martians saw no great changes until the Agrarian (Neolithic) Transformation from hunting and gathering to agriculture 10,000 years ago.

From an archeologist's perspective, the Paleolithic was not quite so homogeneous. Since Lubbock's time it has been subdivided, with sequencing and timing varying by geography. Specialists now distinguish at least among a lower, middle, and upper Paleolithic. The Lower Paleolithic was the longest (2.5 mya to 200 kya) and included the Oldowan and Achulean traditions discussed in the last chapter, both beginning in Africa and spreading to the Middle East, Europe, and Asia.

The Middle Paleolithic (ca. 300 to 30 kya) is best known as the time of the Neanderthals, our close relatives – not our direct ancestors – who were geographically spread from Europe to Central Asia. They evolved by 300 kya outside of Africa, at least no remains are known from that continent. They lived through severe glacial but also warmer interglacial times. Some hunted mammoths and woolly rhinos, perhaps in sight of the great northern ice sheets. Others enjoyed a more moderate climate, for example at sites known in the Levant.

Neanderthal Culture

Every human society has a *culture*, a term numerously defined in the literature but for present purposes is simply a set of things you learn while participating in a society, which you would not necessarily learn in some other society. The culture includes a language, social customs and rituals, technological know-how, food preferences, and so on. It was once thought that culture was a strictly human phenomenon, but field studies of nonhuman primates show that some also have cultures – very simple ones – that are passed from one generation to another, as when young chimpanzees learn from adults how to strip and shape weeds or twigs to catch termites. Of course, human cultures are much more complex than those of any nonhuman primate, facilitated by our use of language. The essence of modern human culture, however, is not just its complexity but also its astounding variability. Modern people and their customs are different the world over. The people of most societies eat cooked meat, but some societies eat no meat while other people eat raw meat, and a few people eat other people.

Of the hominin species to appear around two million years ago, *Homo erectus* became the most widespread throughout the Old World and is probably ancestral to both Neanderthals and ourselves as well as other species of the genus *Homo*. The culture of *Homo erectus* was more complex than that of any ape. They used fire, constructed dwellings with hearth places, and practiced organized big-game hunting. *Erectus* made a fairly standard Acheulean tool kit (Figure 10.5). From today's perspective, the most puzzling feature of that culture was its constancy. Acheulean handaxes that have been found at sites widely separated in distance, and across 1.5 million years of *Homo erectus* existence, look similar to one another. (Acheulean hand axes are not known from *erectus* sites in China or Java, suggesting that their founders arrived before Acheulean tools were invented.)

The earlier Oldowan stone-tool tradition was also long lived and static. It was as if the appearance of each new grade of hominin brought with it an enhanced sophistication in toolmaking – perhaps in culture generally, but once the new way was in place, it changed little for hundreds of thousands of years until a newly enhanced hominin appeared.

Of all our extinct hominin relatives, we have the clearest cultural picture of the Neanderthals because so many of their remnants have been found since Darwin's day, and they lived primarily in Europe and the Levant, where European paleontologists had easy access to their sites (Wragg Sykes 2020). A common if not universal belief among Western intellectuals, eventually losing ground after World War II, was that groups of living humans had intrinsic differences in quality and intelligence, so what would one think of extinct savages? The originally sketched nineteenth-century picture of Neanderthals was as stooped, ugly, stupid, and brutal cavemen. Ernst Haeckel, Darwin's German disciple, proposed calling similar earlier people *Homo stupidus*, though he did not mean Neanderthals specifically. Such a crude view is illustrated in a 1920 depiction of a Neanderthal family once displayed at the American Museum of Natural History (Figure 11.1).

Now, with increased study and diminution of racist bias, Neanderthals are portrayed more flatteringly, not all that much different from us (Mooallem 2017). Dressed in modern attire and with current hair styles, it is sometimes claimed, they would not stand out riding the subway or in an office (Figure 11.2).

Today's universalist bias in anthropology is seductive but may blind us to real differences. A direct comparison of Neanderthal and modern human skulls suggests that a Neanderthal in a subway or office would draw inordinate attention (Figure 11.3). A more intriguing question, more difficult to answer, is how different were their cultures from cultures of anatomically modern *H. sapiens* 50,000 years ago? Could Neanderthals speak as well as *Homo sapiens*?

Figure 11.1 Charles Knight's 1920 depiction of a Neanderthal family, used in an American Museum of Natural History exhibit.
Source: Wikipedia Commons

Figure 11.2 Reconstructed Neanderthal man in a business suit at the Neanderthal Museum, Mettmann, Germany.
Source: Clemens Vasters

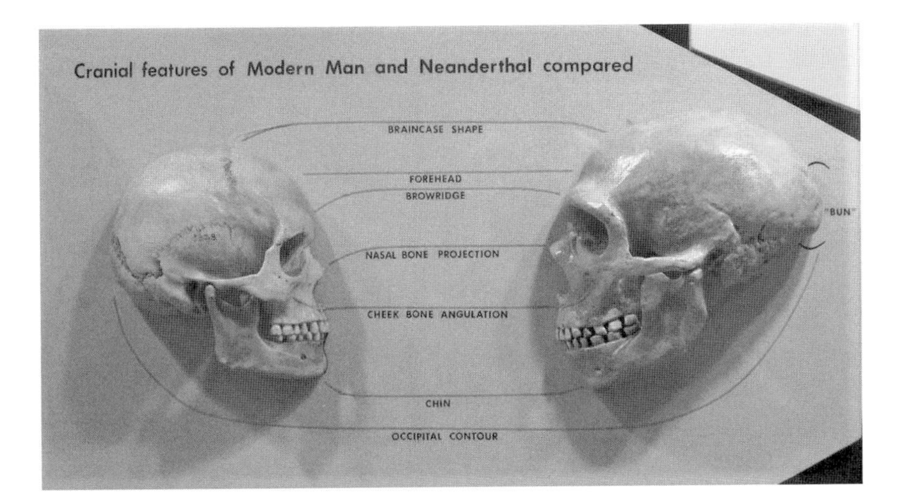

Figure 11.3 Comparison of modern human (left) and Neanderthal skulls, from Cleveland Museum of Natural History. The left cranium is more smoothly globular, the face flat with jutting chin. The right cranium is oblong with a protrusion or "bun" at the rear; there is a thick brow above each eye orbit, a large nasal opening, and a massive jaw with receding chin.
Source: Wikipedia Commons

By now bones from hundreds of Neanderthal individuals, adults, and children have been found, including many skulls and some nearly complete skeletons. Early reconstructions, showing them bandied legged and stoop postured, seem to have been partly due to arthritic distortions of bones. Today it is clear that their skeletal postures were straight like ours, fully capable of striding or running. On average, Neanderthals were slightly shorter but more robustly built then we are, perhaps an adaptation to the colder climate as in stockier bodies of today's Inuit and Eskimos. They were powerfully built and probably could beat us in hand-to-hand combat.

Neanderthals must have worn some kind of clothing to protect against the cold, though we have no tangible remains. Animal furs were an obvious material. There is no sign of eyed needles, as are found among tool kits of contemporary *Homo sapiens*, so Neanderthals lacked the ability to tailor skins into clothes that closely fit the wearer's body to improve efficiency against the cold. We presume they simply draped skins over head and shoulders and around legs and feet, perhaps lacing the pieces together using awls and thongs or simply tying them in place. Sheltered in caves and around fires, this must have sufficed to survive the sometimes-brutal climate. Where caves were unavailable, Neanderthals camped on open ground, these sites indicated by scatters of their characteristic tools and elemental hearths as simple as a pile of ashes, maybe using some sort of tent made from skins and wood or bone.

Perhaps surprisingly, the average Neanderthal brain was slightly larger than that of modern humans, though size alone does not imply they were smarter. Almost surely, they were brighter than their predecessor *Homo erectus* if we judge by their cultural traditions, mostly revealed, as usual, by stone relics. Neanderthal tools and other artifacts were different, more varied, and more sophisticated than relics known from the Lower Paleolithic. Labeled "Mousterian" after the Le Mousier rock shelters in the Dordogne region of France, these cultural remnants, including hand axes, blades, awls, and diverse scrapers, are characteristic of but not exclusive to Neanderthal sites (see Figure 10.6).

A feature of Mousterian industry was its distinctive method of stone knapping known as the *Levallois technique*. Rather than chipping away at a rock to form a single implement, like an Achulean hand axe, a flint lump was preliminarily shaped into a core by striking away superfluous material. This operation left a flat platform on one end that could be struck with a hammer stone, knocking off in one blow a long flake, which needed only to be finished into the desired tool. Many flakes were struck from the same core. We may imagine the core being carried around, its owner striking off flakes as needed. The Levallois technique, first employed in Africa by non-Neanderthal people, was later adopted or reinvented in Europe by Neanderthals (Jordan 1999).

Mousterian tools and weapons were made of stone, usually flint, locally acquired in contrast to the Upper Paleolithic when raw materials would be traded over longer distances. Stone was not the only material used, but it survives while implements of wood, bone or antler rarely do. Moderns can only guess what surviving tools were used for. Microscopic examination of the wear on edges suggests a common use was woodworking. Many objects are simply called "points," some having facets on either side of their base, apparently for hafting, i.e., inserting them into the split end of a wooden pole, held in place by a sticky tar distilled from birch bark or tied with a thong (Kozowyk et al. 2017). Such poles, almost certainly spears, have been found and demonstrated to be effective in penetrating animals at close quarters, however they are not well suited for throwing a distance with force. Neanderthal hunters must have stealthily approached their larger prey before lunging with the spear. This may be one reason why Neanderthal bones show many fractures. Some incapacitating injuries are well healed, indicating that caretakers saw to the injured person.

Primarily meat eaters, if only because edible vegetation is scarce in freezing weather, hunting or scavenging would have been a primary activity for Neanderthals. Small animals could be taken by a solitary hunter but larger or more dangerous ones, including mammoths, rhinos, or cave bears, must have required cooperating groups. Virtually every part of the animal was useful, including meat, fur or hairy skin, teeth, organs, bones, and tusks. It is tempting to think of a wolf pack rather than a troop of apes as the model for a cooperative hunt (especially if we doubt the hunters' language ability). Some animals were trapped before being dispatched, or a herd spooked to run off a cliff, the bodies then collected below. Neanderthals could exploit seashore habitats, gathering shellfish, crabs, and fish from the ocean as well as waterfowl, though apparently not as commonly as modern humans did. Limited Neanderthal seashore sites lack the shell beads and shell containers for paint, found at modern human sites (Will 2020). Finds of human bones with cut marks suggest they too were butchered, but it is not known if cannibalism was frequent or rare, or whether for ritual purposes or in periods of food scarcity, or to celebrate victory over an enemy. Whatever the food, Neanderthals were capable of a range of cooking methods including roasting and smoking.

Many claims of Neanderthal art, ornamentation, adornment, and aesthetic structures have been made, as well as ritualistic burials. Typically, the evidence for these claims is controversial (Pomeroy et al. 2020). If true, they would show Neanderthals capable of symbolic thought and cognitively comparable to modern humans.

Anatomically modern *Homo sapiens*, resident in North Africa by 200,000 years ago, made a noticeable move into the Levant by 50,000 years ago, probably not for the first time, and subsequently through Europe and West Asia. Still living as hunters and gatherers, we assume they had our own level of cognitive and linguistic abilities and, for the times, an unusually sophisticated culture, judging by their relics. There they lived in proximity to Neanderthals for several thousand years, also having contact with Denisovans, at least to the extent of sharing genes.

Novelists have played with the notion of an Ice Age encounter between modern Cro-Magnons and Neanderthals, most famously in William Golding's *The Inheritors* (1955) and Jean Auel's *Clan of the Cave Bear* (1980). Both are premised on one group soon to be replaced by the other. Both plots include an orphan child of one kind adopted by a band of the others, and eventually the act or inference of interbreeding. Of course, the authors did not know their sexual fancies would today be verified by genetic facts.

The Upper Paleolithic "Creative Explosion"

Don Marcelino de Sautuola (1831–1888), a nobleman and amateur archeologist, had an estate in northern Spain where he occasionally dug for artifacts in the Altamira cave. The story that comes down to us, whether or not completely accurate, is that his young daughter Maria accompanied him one summer day in 1879, idly walking into one of the cave's side chambers. While her father was searching the ground, Maria looked upward to the ceiling and saw what no one else had seen in modern times. Illuminated by her lantern were large paintings of animals in vivid color. Mostly bison, but also deer, boar, and horses, some six feet long, skillfully drawn in various poses. In what is now known as the Great Hall, its 1,800-square-foot ceiling has some twenty-five paintings, most in good condition, as well as three or four times as many faded paintings, and many engravings.

In later days, father and daughter explored the cave, finding one decorated gallery after another. Many paintings looked fresh and clear, preserved in the nearly sealed chambers with constant temperature and humidity. As word got out, appreciative visitors came to the site, including the king of Spain, Alfonso XII, with his entourage.

The next year, when Don Marcelino and a supportive paleontologist from the University of Madrid displayed copies of the paintings at the International Congress of Anthropology and Prehistoric Archeology in Lisbon, they were first challenged and then ridiculed. These could not be prehistoric; they were too fresh and bright and skillfully rendered. One French expert, Emile Cartailhac, was scathing and like other delegates to the Congress, refused even to visit the site, insisting they were of recent provenance, perhaps painted with intent to deceive.

Over the next decades, other cave paintings were found of undoubtable veracity because slowly accumulating carbonate deposits covered some of the pictures. Prehistoric dates for Altamira became more plausible, opposition abated, then turned toward affirmation. In 1902, Emile Cartailhac published his *mea culpa* and finally viewed the site with Maria, but by then Don Marcelino was dead, his reputation tarnished (Pfeiffer 1982).

No wonder experts were initially dismissive. Small art pieces were known from Ice-Age sites, but nothing on this scale. It was inconceivable to them that primitive minds could produce such expansive masterworks. Nearly 350 caves with prehistoric art have since been discovered in France and Spain, their images not only in large chambers but in some places barely accessible to the painter. The oldest known cave painting is a red hand stencil in Spain, dated to 64,000 years ago, so early that perhaps it was made by a Neanderthal (Hoffmann et al. 2018). But for the most part, cave paintings were created by anatomically modern humans who had moved from Africa into Europe and Asia by 50 kya, if not before.

The earliest known European clearly-figurative rock art, dated to 30 kya and among the best preserved, is in France's Chauvet Cave. One of the most awesome of sites, Chauvet was not discovered until 1994 and is now closed to protect its contents, but its images can be seen in Werner Herzog's documentary film, *Cave of Forgotten Dreams* (2010).

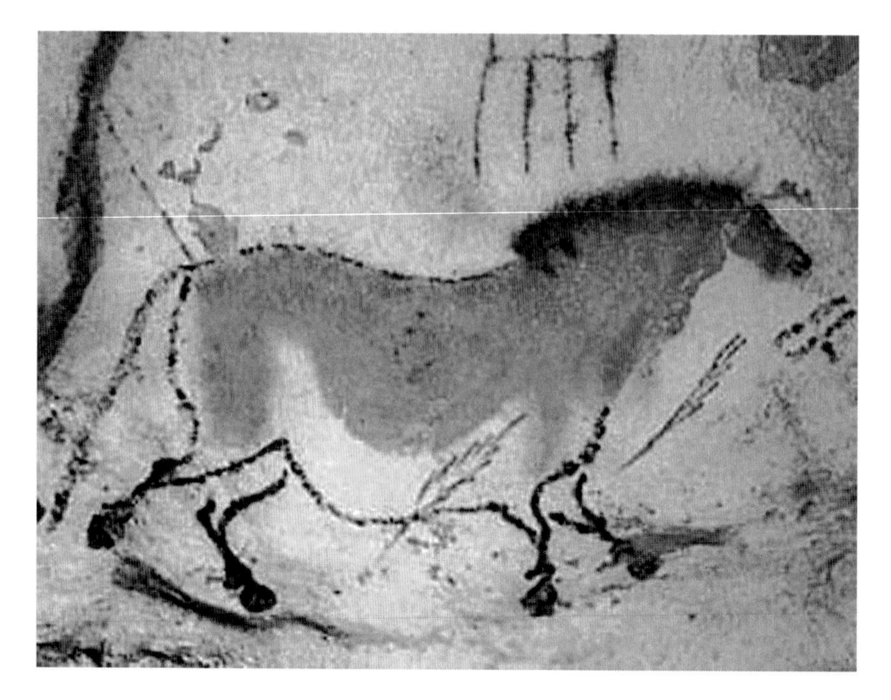

Figure 11.4 A horse (*Equus*) from the Lascaux cave in France, in the original colored tan.
Source: Wikipedia Commons

The oldest known animal images are outside Europe, a bull and a warty pig in Indonesian caves, dated to at least 45 kya. Thus, in the Upper Paleolithic, humans were producing rock art at opposite ends of Pleistocene Eurasia (Aubert el al. 2014; Brumm et al. 2021).

Decades ago while in graduate school, I hung in my apartment, for no reason other than visual appeal, a picture of a horse reproduced from the Lascaux cave in France, today dated to roughly 17,000 years old (Figure 11.4). Lascaux was discovered in 1940 by four boys, playing in the woods. Coming across a small hole in the ground where a tree had been uprooted by a storm, they widened the hole and slide down into a huge chamber 100 meters long, adorned with paintings. The enterprising youngsters started charging 40 centimes admission. As visitors increased, so did the price, up to two francs. Soon an agent for the aristocratic family that owned the land showed up and demanded half the fees, a week later taking over the cave and developing it into a lucrative tourist attraction, reaching a peak of 125,000 visitors per year by the 1950s (Pfeiffer 1982). But at Lascaux and other caves then open, visitation brought moisture and pollution to the interior, damaging the art, some of it irreparably. All of these caves are now closed to the public. At Lascaux and other of the most famous sites, duplicate caves have been constructed for visitors to tour.

Such paintings were the most stunning innovations in what had previously been a humdrum monotony to the Stone Age, collectively marking a new phase, the Upper Paleolithic, or as it has been appropriately nicknamed, "the Creative Explosion." It ran from about 50 to 10 kya, through the last glacial maximum until the end of the Ice Age. This was the domain of anatomically modern *Homo sapiens*. All forms of archaic *Homo* were by then extinct or would soon be gone or isolated in their last refuges. Rock art, essentially an activity of Upper Paleolithic humans, is known from all inhabited continents, sometimes in caves where it can be exquisitely preserved, sometimes on surfaces exposed to the elements.

European archeologists by the late nineteenth century could already see, from Upper Paleolithic sites to which they had access, that this was a period of "quickening" cultural variation and sophistication, if still a world of hunting and gathering. As always, stone implements were the most durable and easily found artifacts, and even in less explored regions, Africa and East Asia, they showed a fineness and sophistication not seen from earlier times.

Wall art is found on all inhabited continents, in diverse environments, its three major themes geometric patterns, humans, and animals, usually large ones that were hunted, including some still alive, if not in that locale (bison, reindeer, deer, ibex, horses) and some extinct (mammoths, aurochs, woolly rhinoceros, cave bears, cave lions). Plants are absent, as are children and babies (Clottes 2011).

Hints of art are tentatively attributed to Neanderthals because their dating seemingly precedes the arrival of *Homo sapiens* in Europe (Hoffmann et al, 2018), but artistry becomes full blown among modern people. Sometimes the pieces are so expertly rendered that the practitioner must have been highly experienced and taught by older artists. Some may be graffiti, drawn by bored youngsters. In any case, the proliferation of elaborately painted caves in France and Spain within a few thousands of years cannot be coincidental. There must have been a shared culture across that region during that period that encouraged the adornment of deep caves and other rock surfaces with realistic images of animals, most of them prey.

Other indications of regionally shared information and ideals are smaller ("portable") works of art including etched bones, painted stones, and small figurines. Among the oldest of these are the figurines that appeared in quantity during the period of increasing glacial advance after 30 kya, most famously the Venus of Willendorf, its type specimen found in 1908 near the town of Willendorf in Austria (Figure 11.5). Many similar pieces have been unearthed across Europe, made of stone or ivory as if from a common template (Mellars 2009). Obviously female, they are obesely pear-shaped with large sagging breasts and buttocks, the hands atop the breasts, and prominently enlarged vulva. They have no facial features or feet, so do not stand upright. About eleven centimeters tall, they are of convenient size to hold in the hand. With nothing else known about them, there have been many conjectures that they were fertility or cultic talismans. Some suggest they are erotic objects, but this seems implausible. As far as we know from historic times, sexiness may be attributed to fatted as well as slim figures but always to youthfully nubile bodies, smooth and

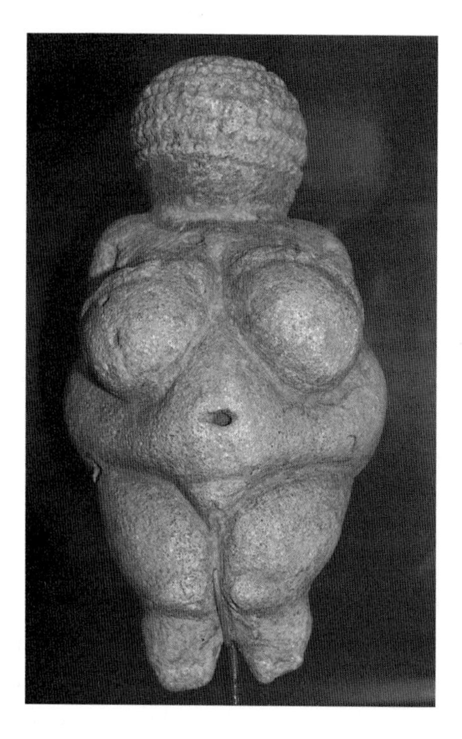

Figure 11.5 Venus of Willendorf, photographed by Don Hitchcock.
Source: Wikipedia Commons

firm rather than flaccid or saggy (Mazur 2009). Admitting the riskiness of mind melding with an Ice-Age man, still it seems too far a stretch to imagine these were objects of sexual desire.

Language and Conversation

Changes in tools and technology were another indicator of the Creative Explosion. The appearance at anatomically modern *H. sapiens* sites of blade industries, in which flakes struck from cores were much longer, thinner and flatter than before and sometimes highly finished, became a marker of the Upper Paleolithic in Europe (though not as important in the Late Stone Age of the Far East or sub-Saharan Africa). There were bone and ivory points, missing from Mousterian assemblages. Also around 35,000 years ago was the introduction of microliths, perhaps a centimeter long, apparently used for spear points and arrowheads. Burins, like chisels or groove cutting tools, were used to work bone, antlers, and hides. Advanced darts and harpoons appear, including light spears with throwing sticks (atlatls) that take advantage of the arm's leverage to achieve greater velocity in the missile, used by 30 kya in Europe, Australia, and later the Americas (McClellan and Dorn 2006). The bow and arrow spread toward the end of the Upper Paleolithic to every inhabited region except Australasia and Oceania. There were spoons, fishhooks, the oil lamp, rope, and the eyed needle.

Among the largest and most enigmatic structures, found across Upper Paleolithic Eastern Europe around the time of the Last Glacial Maximum, are concentrated rings of mammoth bones seven meters in diameter, surrounded by large pits often interpreted as evidence for storage of food or bone fuel, rubbish disposal or simply as quarries for loess used to construct the features (Handwerk 2020; Pryor et al. 2020). The circular structures themselves are widely considered the remains of dwellings that offered shelter during the long, full-glacial winter seasons if not year-round (Shipman 2015).

The dog, first species and only carnivore ever domesticated, diverged from wolves perhaps 20 kya, about the time of the Last Glacial Maximum and during this period became domesticated (Bergstrom, P. et al. 2020). The first undisputed remains of a dog buried with a human dates to 14 kya. It was not until 11 kya that people living in the Near East domesticated aurochs (now extinct, the ancestor of domestic cattle), boar, sheep, and goats (Irving-Pease et al. 2018).

Whereas it had once been suitable to give specific names for simple, static tool cultures lasting over long periods of time and region (Oldowan, Achulean, Mousterian), tools of the Upper Paleolithic were changing so quickly, and so much from one region to another, that new names proliferated, most famously Aurignacian, Gravettian, Solutrean, and Magdalenian, just based on Western Europe. The assumption was too facilely made that each name represented a distinct human culture. Some paleoarcheologists now consider the naming tradition counterproductive (Shea 2017). For present purposes, the main point is that the Upper Paleolithic (50 to 10 kya), province of anatomically modern humans, was unusual in the increasing sophistication and variation in culture, as far as can be discerned by its remnants. Why this change? It cannot simply be that archaic *Homo* like Neanderthals were replaced by anatomically modern *H. sapiens*, because modern sapiens had already been around for a long time, their earliest fossils dated to more than 200 kya. Was some Rubicon crossed since then that accounts for the Creative Explosion of the Upper Paleolithic?

The difference between Neanderthal behavior (indeed, all former human behavior) and modern human behavior during the Upper Paleolithic is amazing. We now see a variety of new technologies, rapidly changing, as never seen before. Ways of life, clothing, burials with undoubted grave goods that infer conceptual belief, construction of housing, personal adornment, the proliferation of portable art and most impressively of magnificently painted caves that on first sight were barely believable to modern eyes, accumulated over the past 50,000 years, most visibly in Europe and the Middle East, but to a degree elsewhere too (Jordan 1999).

The Language Puzzle

Homo sapiens has been suggested to be unique among the primates for largest brain (not counting Neanderthals, whose brains were larger), penis size (no Neanderthal fossil survives for comparison), and body hairlessness (the "naked" ape). It is perennially offered that we are the only species trying to define its uniqueness.

Today there is nearly a consensus that language in the sine qua non for humanity. Apes taught an artificial "language" of symbol manipulation communicate impressively – for apes – but do not seriously challenge the singularity of our own linguistic capability.

Speech is so obviously the method of choice for human communication that it has been construed as essential for language. Debates about the extent of Neanderthal language once focused on the anatomy of its larynx and whether it had sufficient sound-producing capability, which it did (Holdren 1998). But does that really matter? Perhaps Neanderthal "language" was based on gestures rather than speech (Corballis 1999). Language resides in the brain, not the voice box. Signing systems are found in all communities of deaf people, each a distinct and full language with the same sort of grammar found in spoken languages (Senghas, Kita, and Özyürek 2004). Furthermore, the grammar of a true sign language is unrelated to surrounding spoken languages. American Sign Language (ASL), commonly used by the deaf community in the United States, resembles neither spoken English nor British sign language (Pinker 1994).

Chimpanzees in the wild have richly gestured communication based on facial expressions, hand motions, and body postures. From the perspective of the primate series (Chapter 10), we would expect early hominins to have used this kind of communication at a minimum, but when did they begin talking to one another? This was a popular question among nineteenth-century scholars, but since tangible evidence is out of reach, their debates were invariably inconclusive and increasingly boring, to the point where some scholarly groups declared the topic out of bounds.

Students of language found more productive pursuits, one being comparative linguistics, which had impressively demonstrated that most Eurasian languages west of the Himalayas evolved from a common prehistoric root, reconstructed as Proto-Indo-European, spoken in the Neolithic era.

Another line, strongly associated with Noam Chomsky (1928–), argues that basic linguistic structures are wired into human brains, taking full form as the child grows up in a particular language community.

To a sociologist, classic Chomskyan linguistic theory, in its concern for the individual's unconscious knowledge of grammar, seems oddly asocial. Human communication is essentially an interaction. One person talks (or signs) to another, whether to convey information or for information-free social functions such as affiliation, nurturance, seduction, or status attainment. The script may include body postures, facial cues, and hand movements that give added meaning to the exchange of words. Altogether, features of conversation allow considerable communication and functional social interaction even without language. On the other hand, language competence does not imply the capacity to converse. Children with the variant of autism known as Asperger's Syndrome pass the normal linguistic milestones, yet they cannot carry on a normal conversation, having difficulty with eye-to-eye gazing, facial expression, and body postures that regulate the exchange. An Asperger's child might speak articulately about something of interest, then walk away when someone else begins talking, without recognizing their departure as inappropriate. Thus, conversation is a considerably different ability than language per se.

Turn-taking, the exchange of speaker and listener roles, is central to human conversation. This has not been reported among living primates, suggesting that it did not evolve until well along the hominin path toward ourselves. Possibly the enlargement of the hominin brain, which differentiates *Homo* from australopithecines, carried with it some enhanced communication ability. But with no corroborating evidence, it seems we are no closer to finding the beginning of conversational language then were the scholars of the nineteenth century.

There is, however, one additional clue, not conclusive but suggestive. It seems inconceivable that the quickening cultural changes of the Creative Explosion could have occurred, and been transmitted across generations, without language comparable to our own. That cannot be solidly proven, but nearly every writer on this topic concurs. However, that leads to more questions. If anatomically modern *H. sapiens* have been around for over 200 kya, why did the Creative Explosion not occur until 50 kya? Were some important linguistic mutations, or some further brain evolution invisible in the fossil record, introduced midway along the *sapiens* lineage? Is the dating of the Upper Paleolithic off, perhaps biased by particular archeological sites in western Eurasia that have been explored? Were *Homo sapiens* always neurologically capable of language, but required 150,000 years to build up a sufficient cultural substrate upon which that capability would become manifest? We are back to intractable questions, perhaps best put aside for another century, when we may be better able to answer them. Whatever the reason, it is clear that by the end of the Pleistocene Ice Age, humans were showing impressive talents as artists, tool makers, and big game hunters.

12 EXTINCTION OF LARGE ICE AGE MAMMALS

Thomas Jefferson was an avid naturalist. As ambassador to France, he faced ridicule from French savants, notably the Comte de Buffon (1707–1788), preeminent natural historian of the late eighteenth century, over the paucity of impressive animals in North America. The deer, the beaver, the raccoon all seemed puny compared to charismatic animals of Asia and Africa.

Later, as his nation's third president, Jefferson commissioned Meriwether Lewis and William Clark to explore the Northwest Territory newly acquired in the Louisiana Purchase. He specifically asked them to look for mammoths, knowing their discovery would give him bragging rights over his European friends. Lewis and Clark did find mammoth teeth but no living specimen.

They also met their first grizzly bears, perhaps more impressive, certainly more fearsome than mammoths.* A journal of their 1804–1806 expedition records that "Indian tribes give a very formidable account of the strength and ferocity of this animal; which they never dare to attack but in parties of six, eight or ten persons; and are even then frequently defeated with the loss of one or more of their party." Initially Lewis thought the danger overblown, at least for his better armed men, but by the end of his trip had second thoughts: "These bears being so hard to die rather intimidate us all; I must confess that I do not like the gentlemen and had rather fight two Indians than one bear."

* The grizzly bear (*Ursus arctos*), also known as the American brown bear, is much larger and more dangerous than the far more numerous American black bear (*Ursus americanus*), which to endless confusion is often brown in color.

One harrowing encounter occurred by the Missouri River on May 15, 1805.

> "Six good hunters of the party fired at a Brown ... Bear several times before they killed him, and indeed he had like to have defeated the whole party. He pursued them separately as they fired on him, and was near catching several of them. One he pursued into the water. This bear was large and fat and would weigh about 500 pounds ... He pursued two of [the men] separately so close that they were obliged to throw aside their guns and pouches and throw themselves into the river although the bank was nearly twenty feet perpendicular. So enraged was this animal that he plunged into the river only a few feet behind the second man ..., when one of those who still remained on shore shot him through the head and finally killed him."

By its end, men of the expedition had killed over thirty grizzlies with no losses of their own. Lewis and Clark reported back to Jefferson that the nation indeed had a truly impressive mammal in the wild. (See their journal online at lewis-clark.org. I have corrected original misspellings and punctuation.)

Shortly after Lewis and Clark's return, Captain Zebulon Pike purchased a pair of grizzly cubs as a gift for Jefferson, writing the president that they were considered by the natives the most ferocious of animals. Pike transported the cubs to Washington, where they spent a short period in an enclosure on the White House lawn but soon were dispatched to Peale's Museum in Philadelphia. As the bears matured, they became less cute than threatening. After one broke out of its cage and terrorized the Peale family, it was shot and the other put down as well. Both were mounted for display at the museum (Wilson 2008).

California's state flag features a grizzly bear, but it no longer lives there, having been hunted to extinction. Nor is it found in most of the lower 48 states, except around Montana and Wyoming, near Yellowstone and Glacier national parks, where they are protected as an endangered species. Grizzlies are more numerous in British Columbia and Alaska.

Believing in Extinction

During the past 50,000 years, large mammals lived in America and nearly everywhere else that was habitable, including many islands.

By 10,000 years ago, most were extinct. The past 50,000 years are often called "Near Time," approximately the duration over which radiocarbon dating is trustworthy.

Thomas Jefferson did not think extinction could happen. Though a Deist, he believed in the medieval concept of a Great Chain of Being, a hierarchical structure devised by God to put all creatures in their place, from the lowest up to Man, who ranked just below the angels. It was implausible to Jefferson, and many others of his day, that one link in that God-given chain could cease to exist. But by the mid-nineteenth century, the fact of extinction was widely acknowledged. With so many remains of huge and unfamiliar animals being dug up, it was implausible they could still be alive but unnoticed.

In the late 1700s, the bones of an enormous mammal were found in Argentina. From illustrations of the reassembled skeleton, it was identified by comparative anatomist Georges Cuvier as an elephant-sized sloth and named *Megatherium americanum*. When Darwin was in Argentina, he found another specimen of *Megatherium*, later displayed in London (Figure 12.1). This creature could hardly have remained unseen if it were still alive.

Jefferson himself wrote of a remarkably large but unknown animal found in a West Virginia cave, later identified as a sloth (*Megalonyx jeffersonii*). Smaller than *Megatherium*, it was far larger than living sloths of Central and South America. Cuvier identified mammoth and mastodon fossils (now distinguished as separate species) as extinct relatives of the elephants still living in Africa and India.

Extinctions in Near Time

In 1876, Alfred Russel Wallace, surveying then recognized extinctions, suggested

> ... we are in an altogether exceptional period of the earth's history. We live in a zoologically impoverished world, from which all the hugest, and fiercest, and strangest forms have recently disappeared ... Yet it is surely a marvelous fact, and one that has hardly been sufficiently dwelt upon, this sudden dying out of so many large mammalia, not in one place only but over half the surface of the globe. (quoted in MacPhee 2019)

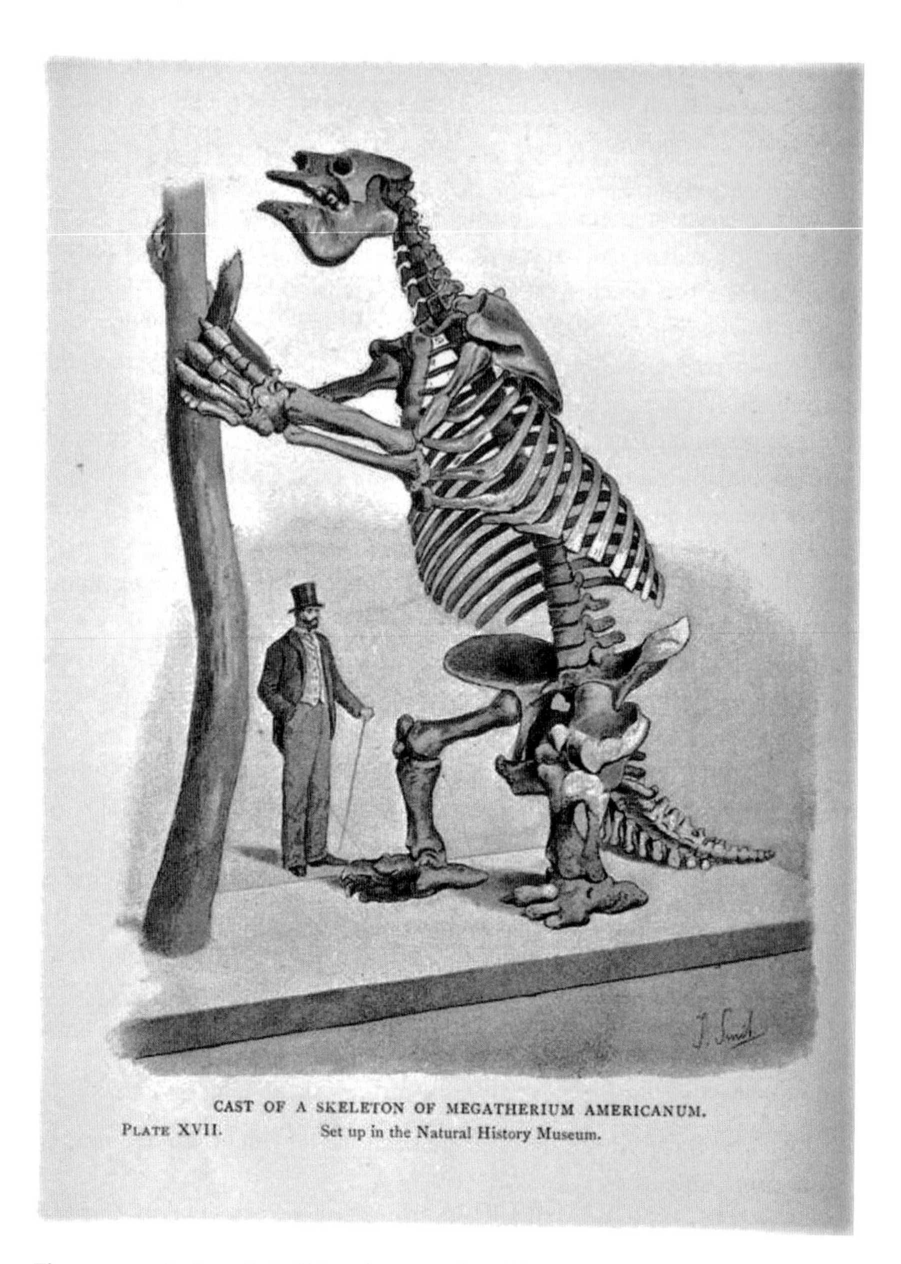

Figure 12.1 A giant sloth (*Megatherium*), found by Charles Darwin in Argentina and displayed in the Natural History Museum of London.
Source: Wikipedia

Wallace's observation has since been abundantly reconfirmed. During the late Pleistocene and afterward, many large and strange-looking mammals disappeared either totally or from continents where they had long thrived.

These so-called megafauna, mostly mammals, are often defined as exceeding about 45 kg (100 pounds) in adult weight, though some were far heavier. They are charismatic creatures, holding our interest not only because of their large size but also because, as shown by modern dating methods including radiocarbon dating, so many disappeared so quickly. Large mammals of the Pleistocene that still survive are mainly in tropical Africa and Asia. In North America, all that remain are bison, black and brown bears, cougars and jaguars, caribou, moose, elk and deer, muskox, mountain sheep and goats, wolves, and pronghorns.

We know today that extinctions have always been part of the history of life, sometimes in episodes so great that they are dubbed "mass extinctions." The Age of Dinosaurs ended sixty-six million years ago when an asteroid or comet impacted Earth, leaving a large crater off Mexico's Yucatan coast. But apart from such extreme events, there is a routine background level of extinction with species occasionally disappearing from (or newly appearing in) the fossil record. Before that asteroid struck, many kinds of dinosaurs had appeared, then disappeared, throughout the Mesozoic era. Every child's favorite, *Tyrannosaurus rex*, was a late arrival, leaving its fossils long after many other dinosaurs had come and gone. Indeed, critics of the asteroid theory of extinction point out that dinosaurs, overall, were in decline before any extraterrestrial coup de grâce. Trilobites, those "beetles" of the Paleozoic, have a similar story, with genera and species regularly appearing and then disappearing from the fossil record, until the last of them perished at the boundary between the Paleozoic and Mesozoic eras, the largest of all extinctions.

The notion of a *mass extinction* began in popular literature and was taken up by professional paleontologists in the 1980s, who enumerated five such episodes. But just what deserves to be called a "mass extinction" is somewhat arbitrary as they are not easily differentiated, or necessarily more destructive, than other episodes of extinction that occurred more frequently or that failed to qualify, by whatever means of tabulation, as breaking into the top five (MacLeod 2013). The extinction of megafauna near the end of the Pleistocene, certainly higher than the background level during the last Ice Age, is not regarded a mass extinction because it was relatively modest and basically concentrated in two clades, mammals and birds, with some reptiles.

It is likely there were once mammals whose fossils have not been found and identified. Some are known only from traces, perhaps a tooth, while other mammals left plentiful fossils, even entire skeletons. A fossil in hand cannot be assumed to be either the earliest or latest specimen during a species' time on Earth. A well-dated fossil can tell us only that the animal was alive at that time but not how long its kind had been around or how much longer it would be. In North America (above the Tropic of Cancer) fossils of the armadillo-like southern pampatheres (*Pamaptherium*) are known from only two sites, fossils of the modern horse (*Equus*) are known from well over 100 sites (Grayson 2016). There is inevitable uncertainty about what animals were here, where they lived, how big they were, when they first appeared, and when they disappeared. Still, our picture of Near Time extinctions is clear enough to show characteristics not typical of other concentrated periods of extinction:

1. The animals lost were exclusively terrestrial vertebrates. Marine vertebrates largely escaped extinction during Near Time (at least until recently).
2. The biggest species were most affected. Size mattered a lot, likely because most large mammals have low reproduction rates and slow maturation. If we include birds and reptiles that went extinct during Near Time, they too tended to be large.
3. There was a peculiar distribution of extinctions over time and space. They occurred on most continents and many islands, but not at the same time or to the same extent. They could not have had one simultaneous cause like an asteroid impact.

The tallies are usually made in terms of genera rather than species because it simplifies the number of names/kinds of entities one must deal with, and taxonomists are thought more likely to get larger categories correct. Whether counting genera or species does not alter the overall picture. Continents suffering the greatest losses were Australia and the Americas, less in Eurasia, and least in Africa. North America lost thirty-two genera of mammals that weighted more than 40 kg, including mammoths, mastodons, giant bison, several species of horses, camels, deer, peccaries, tapirs, giant beavers, ground sloths, glyptodonts, and pampatheres, and most of its large predators: giant short-faced bears, saber-toothed cats, Ice-Age lions, the American cheetah, and dire wolves. Losses in South America were more severe, forty-seven large mammalian genera, including not only immigrant groups like mastodon-like gomphotheres, horses, saber-toothed cats, and giant bears, but also many of the continent's native giants, including several genera of ground sloths, pampatheres, and glyptodonts (Prothero 2017). During the Pleistocene, the Americas as a whole supported 85 genera of mammals weighing over 100 pounds. Of these, paleontologists think about 65 (three-quarters) did not survive.

Extinctions in Eurasia were less severe. Europe lost only seven of twenty-three genera (mainly mammoths, rhinos, hippos, giant Ice Age deer and cattle, cave lions, cave bears, and scimitar cats). Sub-Saharan Africa escaped with its megamammals largely unscathed; only two of forty-four genera were gone by the end of the Pleistocene.

Mammals

Mammals are warm-blooded vertebrates which in females produce milk for feeding their young, have a neocortex in the brain, body fur or hair, and three middle ear bones. These characteristics distinguish mammals from reptiles and birds. The first true mammals lived in the early Mesozoic Era when most were about the size of a rat with a shrew-like body. In a world dominated by dinosaurs, they probably lived in the undergrowth or underground to avoid predators. The demise of the last nonavian dinosaurs 66 million years ago, along with 75 percent of all other animal species and about half of plants, was one of the greatest – and probably quickest – of mass extinctions.

Paleontologists have been very lucky to find in Colorado datable rock strata deposited conformably during the first million years after the impact. From embedded fossils and pollen, they have reconstructed the remarkably rapid recovery of plant and animal life. Ferns were among the first returning vegetation (perhaps because spores survived better than pollen or fruiting parts), then palms, then legumes. New mammals evolved to take advantage of the nutritious seeds and lack of competition, increasing in diversity and size (Lyson, Miller et al. 2019). In another 40 million years, the body weight of some mammals had reached 10,000 kg (22,000 pounds) (Prothero 2017).

Nineteenth century fossil hunter Edward Drinker Cope (see Chapter 2), having collected many large dinosaurs and mammals, noted a trend for species in some lineages to grow larger over evolutionary time. Perhaps this view was inescapable as diggers of his time uncovered ever larger creatures. This notion of increasing size was dubbed "Cope's law" and is borne out in some cases. The earliest dinosaurs were the size of chickens and some eventually evolved into behemoths. Some mammalian lineages grew from rat-sized to immensity during the Cenozoic. Inevitably, there are structural limits to growth, not least the demands on the skeleton and muscles to support and move an enormous weight. (Gravity is less constraining for aquatic animals. Today's blue whale is the largest animal that has ever lived.)

Why would animals become bigger? Great size can sometimes improve fitness for survival and reproduction. It may be an advantage in competition among males for mating opportunities, avoidance of predators or better capability to defend against them, better adaptability to climate change, better thermal efficiency, supporting larger brains and longer life spans. But in other circumstances, largeness can be a disadvantage, requiring more food and water intake, making the animal more conspicuous and attractive as prey, usually increasing gestation time to produce fewer births.

Cope's law is today recognized as an overgeneralization. Some lineages do not increase in size. Some species restricted to islands or other constrained environments, if they are free from predators and food is limited, may do better if they grow smaller. Many such "dwarf species" are known, essentially miniature versions of animals that are much larger when found in less constrained settings. The most famous of these is an extinct human, *Homo floresiensis*, nicknamed "the hobbit," whose fossils were discovered in 2003 on the island of Flores in Indonesia. Adults stood little over a meter tall (Sutikna et al. 2016).

By 30 million years ago, Antarctica was cooling. The other southern continental plates had separated from it and moved northward, nearing their present positions, each now surrounded by ocean. India was plowing under Asia, pushing up the Himalayas. The two Americas remained unconnected until the Panama land bridge consolidated, forming a migration route by 2–3 mya over which South American natives moved north, including ground sloths, armadillo-like creatures, native hoofed mammals, capybaras, porcupines, opossums, and giant predatory birds. The majority of migrants through Panama were North American natives heading south, including horses, llamas, tapirs, peccaries, bears, cats, raccoons and coatis, dogs, and deer (Prothero 2017).

Habitat

Louis Agassiz wrote of ice sheets as being devoid of all life. But despite the desolate look of a glacial landscape, it does support life. On the glacial surface or in the snowpack, or where ice meets bedrock, are bacteria, algae, fungi, viruses, and rotifers (Hodson et al. 2008; Williamson et al. 2019). There are tiny but visible animals including glacial midges (*Belgica antarctica)*, Antarctica's only endemic insect, freezing and thawing with the seasons. At about five mm long, *Belgica* is the largest purely terrestrial animal native to the continent (Baust and Lee 1981). Other miniature indigenes are snow fleas, ice worms, mites, ticks, springtails, and tardigrades ("water bears"). Copepods (crustaceans), nematodes (round worms), and fairy shrimp, all freshwater invertebrates less than 2 cm long, are active in summer, living in melt ponds or lakes.

Marine animals are plentiful offshore, from very small krill to very large humpback and killer whales, as well as fish and squid. Elephant and leopard seals use both the sea and the shore (www.units.miamioh.edu/cryolab/education/antarcticbestiary.htm" www.units.miamioh.edu/cryolab/education/antarcticbestiary.htm). The southern continent has 46 species of birds including penguins, albatrosses, petrels, cormorants, bitterns, herons and egrets, ducks, geese and swans, gulls, and terns (www.wildfoottravel.com/antarctica/information/wildlife-plants/birds" www.wildfoottravel.com/antarctica/information/wildlife-plants/birds).

Just south of the great Pleistocene ice sheets of North America and Eurasia were belts of unglaciated land, snow covered in winter but exposed tundra in summer. Vegetation in boggy tundra soil is mostly grasses, sedges, dwarf shrubs, mosses, and lichens, all low growing. There were also drier belts of land, the steppes, essentially treeless grasslands. Tundra might change to steppe and vice versa, depending on variations in temperature and precipitation. Still farther south began conifer forests, their margins moving northward or southward with changes in climate.

The vast steppes between the ice sheets and boreal forests, wider in Europe and Asia than North America, provided abundant forage and firm ground, amply supported grazing animals and their predators. Migrating herds of reindeer, horses, and mammoths used the steppes during the short summer, as did saber-toothed cats (Figure 12.2).

Figure 12.2 The image by Mauricio Antón depicts a late Pleistocene landscape in northern Spain with woolly mammoths, equids, a woolly rhinoceros, and European cave lions with a reindeer carcass.
Source: Caitlin Sedwick (Apr. 1, 2008). "What Killed the Woolly Mammoth?" *PLoS Biology* 6 (4): e99. doi: 10.1371/journal.pbio.0060099

Accustomed as we are to looking at the world mapped on a flat sheet with the Atlantic Ocean at its center, we get a sense of Alaska and Siberia as far distant. One glance at a world globe shows them nearly touching, separated by the Bering Strait, only 82 km (51 miles) across at its narrowest point. During the great glaciations, when so much water was frozen into ice sheets that sea level was far lower, the Strait was dry land. "Beringia" is the name given to this former land bridge and adjoining areas, now mostly shallow seas that were then also dry. Lower sea levels widened continental margins and generally increased land surface connections. Much of the expanse from Southeast Asia through Indonesia was then above sea level, although a water crossing was always required to reach Australia.

When Beringia was dry land, animals and eventually humans could walk between Siberia and Alaska. Asian species such as the mammoth, bison, and scimitar cat immigrated to America, while the American fox, ground squirrel, horse, and camel went the other way. Even at the height of the last Ice Age, much of Siberia and interior Alaska were ice free because there was insufficient moisture for snow to accumulate. Oddly from today's perspective, ice-free Alaska was at times functionally part of Asia, being connected to it by Beringia while isolated from America by great Canadian ice sheets to the south and east (O'Neill 2004).

The steppe-tundra habitat, between ice sheet and forest, was enormous in east-west extent, running across Eurasia from the Atlantic to Siberia, over the Beringian land bridge into Alaska and Yukon, and when not blocked by ice reaching below the Laurentide ice sheet (Figure 12.3). Called "the Mammoth Steppe," after its most iconic creature, it was then Earth's most extensive biome, having a cold dry climate, vegetation dominated by palatable high-productivity grasses, forbs and willow shrubs, animals dominated by bison, horse, and woolly mammoth. It thrived for about 100,000 years without major changes until the end of the Pleistocene. There were other biomes including grasslands and savanna on most continents, large stretches of desert at temperate latitudes due to generally arid conditions, and small areas of rainforest near the equator, but the Mammoth Steppe is remarkable for its size and abundance of fossils. By some estimates, its animal biomass and plant productivity were similar to today's African savannah (Guthrie 2001). Animals, and eventually humans, living on the Steppe might look north and see the edge of great glaciers.

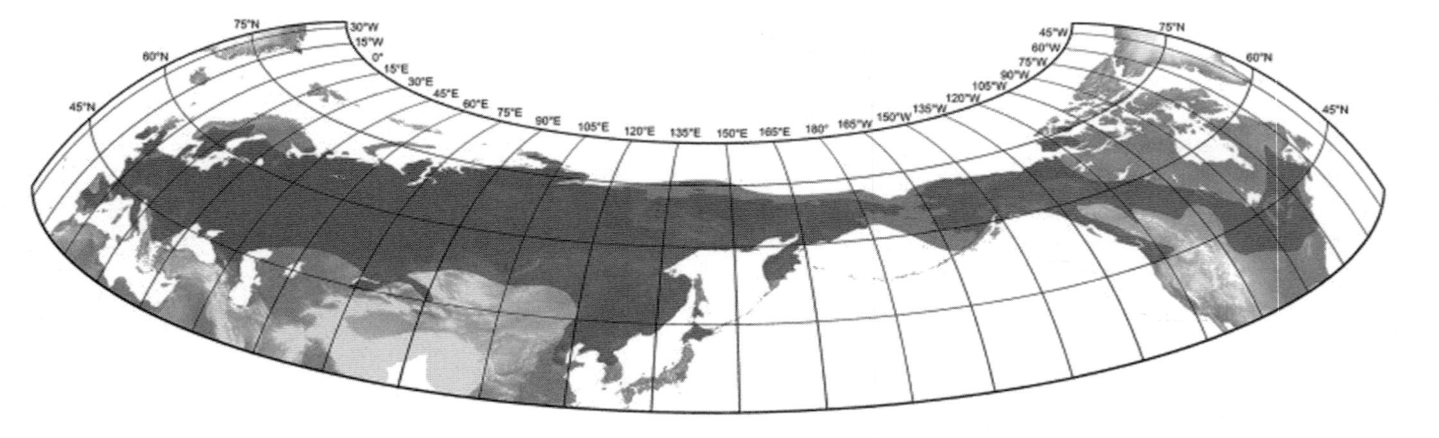

Figure 12.3 Professor Ralf-Dietrich Kahlke from the Senckenberg Research Station for Quaternary Paleontology in Weimar, Germany, has recorded the maximum geographic distribution of the woolly mammoth (*Mammuthus primigenius*) during the most recent Ice Age, based on the current fossil record. Image credit: Ralf-Dietrich Kahlke. doi: 10.1016/j.quaint.2015.03.023

A Brief Sketch of Near-Time Megafauna

Rancho La Brea Tar Pits in downtown Los Angeles is the richest site for fossils of late Ice-Age animals, the oldest dated 38,000 years ago, most between 14,000 and 16,000 years ago. Today surrounded by skyscrapers and freeways, the tar pits were once in dry grassland nourishing more than 420 animal species whose fossils have been recovered from the site. Oil still oozes out of the ground, becoming asphalt as lighter fractions evaporate, at times forming tarry surfaces that can trap animals like flypaper. Predators attacking entrapped prey would themselves drown in the ooze, their flesh decaying and their bones sinking to pit bottoms, preserved from bacterial deterioration (Figure 12.4). Fossils have been extracted at Rancho La Brea for over a century. As its importance became recognized, operations expanded, now including sophisticated laboratories and a museum.

The largest animal found at La Brea is the Columbian mammoth (*Mammuthus columbi*), reaching 13 feet (3.9 m) tall and weighing about 10,000 pounds (4,900 kg). It once roamed throughout the southern half of North America. The American mastodon (*Mammut*

Figure 12.4 Sabertoothed cat and dire wolf fight over the carcass of a Columbian mammoth in the La Brea Tar Pits. Artist Bruce Horsfall, frontispiece of Scott 1913

americanum), smaller than mammoths or elephants, was also at La Brea. Extinct ground sloths are there, the largest, Harlan's ground sloth (*Paramylodon harlani*), over 6 feet (1.8 m) tall and weighing about 3,500 pounds (1,600 kg). Other large herbivores were horses, camels, and deer. All of these, as well as tapirs, peccaries, rabbits, mice, and squirrels, were potential victims of the meat eaters.

The majority of bones found at La Brea are from carnivores, some of them hunters and others scavengers, including the dire wolf, coyote, grey fox, and dogs. Over 1,600 dire wolves (*Canis dirus*) have been found there, more than any other single animal. Bones of over a thousand sabertoothed cats (*Smilodon*), the size of a modern-day lion, were found at La Brea. (How they used those long canines is unknown, one theory being that they ripped open the soft underbelly of a victim, another that they pierced the prey's skull.) Another feline predator, taller and heavier than the sabertoothed cat, was the American lion (*Panthera atrox*). There was also the bobcat, and two species of puma. The largest carnivore at La Brea is the short-faced bear (*Arctodus simus*), which had unusually long legs for a bear, perhaps better for chasing prey. Like other bears, it appears to have been an omnivore. Extant black and grizzly bears have also been found in the tar. The smallest carnivores there are weasels, skunks, badgers, ring-tailed cats, and raccoons.

Over 135 bird species have been identified from bones at La Brea, some of the meat eaters including owls, hawks, eagles, and falcons. Ancestors of the living California condor are in the tar, as is an even bigger bird, Merriam's teratorn (*Teratonis merriami*). One human (partial) has been found in La Brea, a young adult female dated at about 10,000 years old (Arnold 2018).

Extinctions in the Americas

Of the 32 genera of large animals lost from North America around the end of the Pleistocene, three remain extant in South America: the capybara (*Hydrochoerus*), largest living rodent in the world; the spectacled bear (*Tremarctos*); and the tapir (*Tapirus*). South America lost 47 genera of its large mammals, none surviving in the North. There are too many species or even genera to describe here, but we get a feel for what was lost by noting extinctions within the higher and more abstract taxa.

Cingulata

Visualize an armadillo crossing a highway in Texas, where it is the state animal. When startled, rather than running, it instinctively jumps straight up, often into a car passing overhead. Think of a Volkswagen Beetle, perhaps the instrument of death. Now visualize an armadillo the size of a Volkswagen Beetle. That would be the *Glyptodon*, approaching 2,000 kg (4,400 pounds). Glyptodonts and other large armadillo-like creatures, primarily South American animals, died out about 10,000 years ago.

Pilosa

This order includes anteaters and sloths, including the extinct giants *Megatherium* (Figure 9.1) and Jefferson's sloth, already mentioned.

Carnivora

FELIDAE: Gone are the sabertoothed cat (*Smilodon fatalis*), scimitar cat (*Homoherium serum*), and the smaller American cheetah or cougar (*Miracinonyx*). So is the American lion (*Panthera leo atrox*), a larger relative of today's African lions, though the genus survives in South America as the jaguar (*Panthera onca*).

CANIDAE: The extinct dire wolf (*Canis dirus*), most frequently found skeleton at La Brea, was about the same size as modern wolves and is perhaps the most familiar carnivore of the Ice Age, after the sabertooth.

URSIDAE: Gone are *Arctodus* in North America and *Arctotherium* in South America, both very large, short-faced bears, so called because the shape of their skulls seems disproportionately short for a bear. America's surviving bears are the brown (grizzly) bear, the black bear, and in South America the spectacled bear (related to *Arctodus*).

Rodentia

Giant beaver (*Castoroides*) weighted up to 125 pounds (57 kg), much larger that the roughly 44 pounds (20 kg) for today's North American beaver (*Castor canadensis*), also capybaras (*Caviidae*).

Perissodactya

Horses (*Equus* and *Hippidion*) evolved in North America perhaps five million years ago, some migrating to South America and others through Beringia to Eurasia, where they were eventually domesticated. Horses became extinct in the Americas about 12,000 years ago and were absent from the Western Hemisphere until returned by Europeans, particularly Spanish conquistadores.

Artiodactyla

Five genera include peccaries (Tayassuidae), camels and llama (Camelidae), deer (Cervidae), pronghorns (Antilocapridae), and muskox (Bovidae). Primitive camels, like horses, first appear in North America, going extinct there near the end of the Pleistocene, after having migrated to Eurasia, where they evolved into the domesticated Bactrian and dromedary camels familiar today. Eventually they were transported back to the Americas by Europeans.

Notoungulata

Known as toxodonts, these huge South American animals slightly resembled rhinos.

Proboscidea

With a trunk and tusks, these all looked roughly like today's elephants. There were three species of Pleistocene mammoths (*Mammuthus*) adapted to cold climate. The last to emerge was the woolly mammoth (*M. primigenius*), ranging widely across the Mammoth Steppe. Tens of thousands of mammoth bones have been found, including entire skeletons, and even preserved corpses have been found frozen in Alaska and Siberia. They appear in European cave art.

Mastodons (*Mammut americanum*) look similar to mammoths but are importantly different in their teeth, suggesting they browsed leaves and branches, while mammoths grazed like today's elephants. Neither mastodons nor mammoths migrated to South America. One "elephant," the gomphothere (*Cuvieronius*), a name barely known to the public, was in the diet of early South Americans, as indicated by remains found at the Monte Verde habitation site on the Chilean coast, dated to about 14,000 years ago.

Other Continental Extinctions in Near Time

Africa nearly escaped the Near Time extinctions of megafauna. Animals that can be seen there in the early twenty-first century are essentially those there during the Pleistocene.

Eurasia had a moderate loss of large animals, not as severe as in the Americas. And these extinctions seem to have occurred over a longer span of time. Large mammals lost from Eurasia include the cave bear (*Ursus spelaeus*), the cave hyena (*Crocuta*), and the woolly and Merck's rhinoceros (*Coelodonta* and *Stephanorhinus*). There were heavy-bodied Asian antelope (*Spirocerus*) and Eurasian hippopotami (*Hippopotamus*). The woolly mammoth (*Mammuthus primigenius*) that ranged across the Mammoth Steppe from Europe to North America went extinct from both continents at roughly the same time, though some survived on island refuges for several thousand years more. The giant tapir (*Tapirus augustus*) of East Asia was perhaps the largest tapir ever. The Irish elk (*Megaloceros giganterus*), which despite its name ranged across Eurasia, was perhaps the largest deer ever, famous for its enormous rack; it seems to have survived until about 7,000 years ago.

As some species became extinct, so too did those that depended on them. Among Eurasia's top predators lost were the scimitar-toothed cat (*Homotherium*), the cave lion (*Panthera spelaea*), and the European leopard (*Panthera pardus*).

Australia was worst hit and earlier, most large fauna going extinct forty to fifty thousand years ago (Saltré et al. 2016). At the time, New Guinea and Australia were connected by a land bridge, so the whole is referred to as "Sahul." It lost 15 of its 16 large mammals, including the rhino-size *Diprotodon*, the largest known marsupial to have ever lived, its skull roughly ten times as large as a human skull. Other marsupials included huge kangaroo-like animals (e.g., *Procoptodon, Metasthenurus*, and *Protemnodon*). The tapir-like *Palorchestes* was similar in size to a horse. The marsupial "lion" *Thylacinus* was carnivorous but not actually a feline. *Zygomaturus*, weighing 1,000 pounds, looked vaguely like a hippopotamus but was another marsupial. *Phascolonus* was a larger version of today's wombat. *Megalibgwilia* was a genus of echidna, much larger than today's echidna and platypus, the only living mammals that lay eggs.

In addition to the mammals, Sahul's extinctions include enormous flightless birds two meters tall (*Genyornis*), depicted in rock art thought to be 40,000 years old (Gunn et al. 2011). There were huge crocodiles; a 6 meter (20 ft) long python called *Wonambi*; and *Megalania*, a monster Komodo dragon exceeding 6–7 meters (20–13 ft) in length.

Did Modern Humans Do It?

There are two major hypotheses, or schools of thought, competing to explain the extinction of megafauna near the end of the last Ice Age. The one best known to the public is that humans hunted these large animals to extinction as our kind moved into new areas where the prey had not yet learned to avoid them. This has become known as the "overkill" theory, connoting that the killing rate was over what the species could normally sustain.

Archeologists tend to prefer climate change as the most important cause, especially as it altered habitats and food availability. Lest we fall into a false dichotomy, the overkill school does not wholly deny an impact of climate change, and the climate school recognizes clear instances where human hands played a role.

Paul Martin (1928–2010) of the University of Arizona most persistently developed the overkill argument, proposing since the 1960s that when human hunters entered North America about 11,000 years ago, they encountered large herbivores including mammoths, horses, and camels, which had not learned to avoid humans and so were easy prey. Hunters chose the larger animals because they provided plenty of meat. Fairly quickly, these animals' numbers fell to the point where populations were no longer sustainable given the low fecundity of large mammals. As large herbivores became scarce, carnivores that depended on them for food, such as the saber-toothed cat and dire wolf, also became extinct. Humans continued migrating into South America, riding their wave of extinctions, leaving relatively few large mammalian species in their wake (Martin 2005).

Carbon dating was new when Martin began his career, so there were not many trustworthy dates for animal extinctions or human migrations. Still, he could discern a pattern: Loss of large mammals was highest where humans had most recently arrived, in the Americas and earlier in Australia. Losses were lower in Eurasia, which had a longer human presence, and least in Africa, where humans had lived as long as there were humans. Prey animals long familiar with human hunters learned to avoid them, Martin argued, but when prey animals were naïve to the human hazard, they would easily fall victim.

For American extinctions, Martin specifically had in mind the Clovis people, then thought to be the first Americans, known for their distinctive and deadly spear points (Figure 12.5), first found near the town of Clovis, NM and dated to about 11,000 years ago but now put about 13,000 years ago. Kill sites had been found containing both Clovis blades and butchered mammoth bones. Martin thought that two-thirds of the large animals of North America suffered rapid extinction, the direct or indirect effect of Clovis hunting. It was the story of the grizzly bear writ large.

To fortify his argument, Martin pointed out that human colonization of previously uninhabited islands was often followed by extinction of vertebrates living there. For example, when people colonized New Zealand, about 750 years ago, there were nine species of huge flightless birds, the moas (order Dinornithiformes), some weighing more than 500 pounds. Within a few hundred years, all were extinct (Anderson 1989). This is a common pattern: extinctions follow the arrival of humans at previously uninhabited islands. But does it apply to continental extinctions?

Martin emphasized the importance of first encounter between humans and large mammals. This appealing prey would have been

Figure 12.5 Clovis points from the Rummells-Maske Site, 13CD15, Cedar County, Iowa. These are from the Iowa Office of the State Archaeologist collection. Attribution: Billwhittaker at English Wikipedia, under the Creative Commons Attribution-Share Alike 3.0 Unported license

innocent to the hazard presented by the hunters and therefore unlikely to take evasive actions. He envisioned this encounter as a lightning strike, a *blitzkrieg*, with humans wiping out the unwitting prey before they could adapt or evolve a defense to the new danger. This, he explained, was why extinctions were so severe in the Americas and Australia, where humans were relatively recent arrivals. Extinction was less severe in Eurasia, where humans were long present, so animals had habituated to them, and nearly absent in Africa, where humans were "always" present. But this *blitzkrieg* picture also introduces a difficulty when the same animals lived in both North America and Eurasia. If the rapid extinction of woolly mammoths in America depended on first encounter with human hunters, then why did Eurasian woolly mammoths, long accustomed to human presence, go extinct at roughly the same time?

The overkill hypothesis, at least as originally stated, has severe difficulties. There is now good evidence for pre-Clovis people in the Americas. Chile's Monte Verde site is assigned dates of 14,000 to 15,000 years ago, one to two thousand years before Clovis culture, and other sites push the earliest occupation back another thousand years (Braje et al. 2017; Waters 2019). Also, the deadly effect of humans is not exclusively via hunting but also by factors collateral to human arrival, including modifications to the environment as by fires and introduction of new diseases and predators. Even on New Zealand, people destroyed habitat by burning the landscape and brought with them, like most ocean voyagers, the Pacific rat (*Rattus exulans*), a disease vector and itself a predator of small animals. Had Martin known it, he might have emphasized the importance of hunting by dogs, which apparently accompanied ancestral Americans in their migrations through Beringia from Siberia (Perri et al. 2021).

Furthermore, Clovis points, which Martin regarded as uniquely lethal, are best seen as examples of the fine weaponry being produced throughout the Upper Paleolithic but not the only American examples. Points from the Western Stemmed Tradition, dated from Idaho at 16,000 years ago, made spear or arrow heads seemingly capable of dispatching large prey (Davis et al. 2019).

Professor Donald Grayson of the University of Washington was a persistent critic of Martin's overkill theory. The two men and nearly all their colleagues agreed that prehistoric colonization of an island was routinely followed by extinction of some vertebrates. They disagreed about whether this occurred on a continental scale.

Grayson accused Martin of circular reasoning, of assuming that megafaunal losses in North America coincided closely in time with the arrival of the first well-armed (Clovis) hunters, essentially assuming his conclusion. Grayson countered, "Even today we can only show that 17 of the 37 (extinct) genera lasted until about 11,000 years ago" (2016: 278). Thus, extinctions may have occurred before the arrival of humans in North America. Martin seemed to evade this important point: "The timing of extinction of about half of the Americas' lost genera remains to be determined critically, however there is no indication that any noted here endured after 13,000 years ago" (Martin 2005: 119).

Grayson further challenged Martin on the grounds that we have no evidence that prehistoric hunters actually preyed on the range of Ice Age mammals that went extinct. Of 15 North American sites (as of 2016) showing evidence of human hunting, scavenging or butchering of extinct mammals, the usual prey was a mammoth, with one or two sites also of gomphothere, mastodon, horse, or camel killing (Grayson 2016: 133). We must weigh this lacuna against the probability of finding a 10,000-year-old kill site, especially for prey whose killing and preparation would be less prolonged and conspicuous than that of a mammoth.

These complications can be accommodated by today's proponents of human causation. It need not have been the Clovis people specifically who were the critical human destroyers, and any human impact on large herbivores need not have been exclusively via hunting. The Monte Verde site, near the southern tip of South America, along with other pre-Clovis sites that are candidates for early human occupation, were seemingly home to capable hunters (Braje et al. 2017; Davis et al. 2019).

Since Martin began his career, far more fossils and sites of interest have been located, with carbon dates that are more trustworthy. However, there will always remain difficulties in judging the simultaneity of human arrival and species extinction. No fossil can be assumed to be the last living specimen of that species, so prehistoric dates of extinction can only be put at some time after the date of the most recently known fossil. A species lost from one area may continue to exist in refuges elsewhere. Woolly mammoths, almost certainly extinct on all continents for about 10,000 years, survived on isolated Wrangel Island, in the Arctic Ocean 150 km (93 mi) off the coast of Siberia, until 4,000 years ago (Vartanyan 1995). Radiocarbon dating shows human habitation on the island by 3,700 years ago, though there is no direct evidence of hunting.

Despite imprecisions in timing, it is clear that earliest dates of known human occupation were not necessarily followed quickly by extinctions. Humans coexisted for thousands of years with large animals that would later become extinct. Martin assumed that human hunters would go immediately for any vulnerable source of meat, but we know there are cultural modifiers to any such tendency. From today's cultural landscape it is obvious that people differ in their preferences or taboos regarding what is edible or desirable, what can or cannot be hunted (or gathered), not only for food but other uses. A human group, to be truly destructive as hunters, must have the techniques and type of weapons that together efficiently dispatch large animals. No one thinks Neanderthal hunters wiped out any of their prey species if only because Neanderthals lacked the hunting effectiveness of modern humans.

A recent extinction, fairly well documented, is the American passenger pigeon (*Ectopistes migratorius*), which in the early nineteenth century was so numerous that passing flocks were said to have darkened the sky. John James Audubon estimated over a billion birds in a single flock. Martha, the last passenger pigeon, died in 1914 at the Cincinnati Zoo. Overhunting is often blamed, but it seems implausible that there were so many hunters firing so many bullets that they could bag more than a billion birds. Conversion of wild habitat for agriculture may have been as important a destroyer. In any case, humans must shoulder the blame for that particular loss. But extinction did not begin immediately after the first European arrival in the fifteenth century, not to mention the far longer presence of Native Americans. One must allow time for the Europeans to multiply in number, for them to fancy pigeons as targets, and for them to have invented and adopted rifles with sufficient accuracy to shoot pigeons in great number, or to put sufficient acreage under the plow, to produce the extinction. There should be no expectation that an extinction quickly followed the first arrival of *Homo sapiens*.

Great Plains bison nearly were hunted to extinction by sportsmen piling up corpses as they gained easy access to the middle of the continent on the new railroads. This comes close to an example of Martin-style overkill, but in fact, the American bison still survives, if in far diminished herds.

Mammals need not be innocent of human presence to allow close approach. In my own city, deer have become so profuse and unbothered by humans (except when struck by a car) that they congregate in streets and yards, seemingly blasé about people passing by, still watchful but not bolting. Ice Age hunters may not have had only a single "first encounter" for large mammals to be approachable prey. On the other hand, mammals can quickly learn to avoid or outsmart human traps (R. Mazur 2015).

Environmental Causes

If humans were not directly or indirectly the most important cause for extinction, the alternative must be some sort of environmental disruption, the most favored being climate change. Long-term variations in regional temperature and aridity affect the livability of habitat, especially the availability of good grazing or browsing, and the drying up of freshwater sources. Animals capable of migrating to better pickings would have been alright, but those with limited mobility, for whatever reason, would have starved. Very likely the isolated woolly mammoths of St. Paul Island off Alaska, which survived nearly as long as those of Wrangel Island, finally succumbed about 5,000 years ago as sea-level rose and a drier climate produced freshwater scarcity (Graham et al. 2016).

The hypothesis that climate change was the major cause of continental (as opposed to local) extinctions has its own problems, most obviously that despite the enormous changes in global climate that occurred between the dozens of glacial and interglacial periods, none is associated with a spate of extinctions. During the last glacial maximum, between 20,000 and 26,000 years ago, the world as a whole was not only cooler but much drier, as indicated by the vast expansion of desert, grassland, and steppe environments on all continents except Antarctica. Humid tropical forests retreated to small, discontinuous areas near the equator. Yet there were few vertebrate extinctions at that time (MacPhee 2019). The Australian extinctions occurred mostly during a relatively mild period 20,000 years before the Last Glacial Maximum. Perhaps the Younger Dryas, that last episode of great freeze as the Ice Age was ending, was unusually lethal because of its swift onset and severity, but it could not have caused the great extinctions in Australia, which occurred thirty thousand years earlier, and why would it have been so devastating to North and South America megafauna but not to those of Eurasia or Africa or on island habitats? Island extinctions all occurred several thousand years after the Younger Dryas (Burney and Flannery 2005).

Another problem is that climatic hypotheses would predict many extinctions among plants, usually more affected by climate change than animals, but there were no waves of vegetation extinction. Also, climate hypotheses would predict that small warm-blooded animals would be more affected than large ones because smaller animals have higher surface-to-volume ratios and are thus less efficient in holding their body temperature constant, but it was the larger mammals that suffered most (Fernandez 2015).

Recently there have been attempts to statistically compare the importance of anthropogenic verses climatic causes of Ice Age extinctions. In their analysis of 177 large ($\geq$10 kg) mammal species that appear to have gone globally or continentally extinct between the beginning of the last interglacial period (132,000 years ago) and one millennium ago, Sandom et al. (2014) tallied 18 extinctions in Africa, 38 in Asia, 26 in Australasia, 19 in Europe, 43 in North America, and 62 in South America. Focusing on species rather than broader classifications, and breaking geography down to national boundaries, their statistical analysis had unprecedented resolution. Their results showed the severity of extinction to be strongly tied to human presence, with at most a weak Eurasia-specific link to climate change. This implicates modern humans as the primary driver of the worldwide megafauna losses between 132,000 and 1,000 years ago. Southern South America, southeast North America, western Europe and southern Australia were extinction hotspots, with sub-Saharan Africa and Southern Asia as notable cold spots.

Another statistical analysis focused on extinctions in Australia, which have been especially controversial. Saltré et al. (2015) coupled 659 Australian megafauna fossil ages with high-resolution climate records, showing that extinctions were broadly synchronous across genera and independent of climate aridity and variability over the past 120,000 years. These results reject climate variability as the primary driver of megafauna extinctions, instead estimating that the megafauna disappeared across Australia about 13,500 years after human arrival.

Such results add weight to the argument that humans caused Ice Age extinctions but are by no means decisive. It is a truism among statisticians that results coming out of an analysis are no stronger than the quality of data going into the analysis, and these data are rife with uncertainties.

Much of this uncertainty would be removed if we knew more exactly when the various kinds of megafauna went extinct, and for how long they had previously flourished. Taking North America as an example, 37 genera of large mammals are known lost during the Ice

Age. Thirteen of these have never been reliably radiocarbon dated. Of the 24 genera with trustworthy dates, 17 are estimated to have expired 12,000 to 10,000 years ago. Latest trustworthy dates for the others go back as far as 38,000 years ago, which may reflect a scarcity of fossils (Grayson 2016: Table 4.2).

In the early days of radiocarbon dating, some of these genera appeared to survive longer, but these recent dates have not been replicated and are now considered faulty measurements. This leaves open the possibility that a newly discovered fossil of, say, a mammoth will show that it survived much longer in North America. However, such an unexpected finding is generally discounted, even by critics of human causation, because paleontologists have excavated so many sites in North America, without finding any "extinct" large mammal more recent than 10,000 years, that the absence of evidence is regarded as evidence of absence.

Still, there is no denying the insufficiency of data to settle this issue, and we should leave open the possibility of multiple causes, perhaps specific to each part of the world, or a new explanation altogether. A recent proposal, reminiscent of the asteroid explanation for the demise of dinosaurs, is that some smaller but still significant collision triggered the Younger Dryas and thereby the near time extinctions (Kennett et al. 2015). This has problems of its own, including why it left smaller animals and sea life unaffected, and its inability to explain the extent or different timing of extinctions on each continent. On the other hand, the recent discovery of a 31-km-wide impact crater, beneath a kilometer of Greenland ice, roughly dated to the Pleistocene, leaves one wondering (Kjaer et al. 2018).

13 AGRARIAN TRANSFORMATION

The last great ice sheets reached their farthest extent, the so-called Last Glacial Maximum (LGM), between 26 and 20 kya, then began their retreat. During the LGM, anatomically modern *Homo sapiens*, Earth's sole remaining humans, were still hunters and gatherers but with their technical advances, including tailored clothing, coped well with the cold, when necessary, migrating to more clement areas. The Pleistocene epoch ended 11.7 kya, according to the International Commission on Stratigraphy, a date also marking the beginning of the Holocene epoch in which we now live, its climate warmer, generally wetter with higher sea levels, and less variable. Soon humans would become farmers and herders.

From Pleistocene to Holocene

During the early Holocene, some locations in the Middle East had abundant vegetation due to the warmer, wetter climate and increasing CO_2 in the atmosphere, which supports plant growth. The Levant became hospitable to human settlement, probably at first for part of the year but eventually permanently. Wild vegetation included eight Neolithic "founder crops" that would be important in early agriculture: wild progenitors to emmer wheat, einkorn, barley, flax, chickpea, pea, lentil, and bitter vetch, plus four of the five most important species of domesticated animals: cows, goats, sheep, and pigs. The horse, a fifth species, lived nearby but was not domesticated until much later in central Asia (Diamond 1997).

In college I was taught that agriculture began about 10,000 years ago in the Fertile Crescent, where humans first settled in small communities, cultivating nearby fields. Others in the area lived as nomadic herders, trading their animal products with the settled cultivators, to their mutual benefit. The agrarian lifestyle diffused outward, eventually across the inhabited world. Agriculture and permanent settlement were like chicken and egg, impossible to tell which came first because each required the other.

The "Fertile Crescent" was the center of attention, a name popularized in the early twentieth century by archeologist James Breasted (1916), referring to an arch of elevated land from the eastern Mediterranean coast (the Levant), eastward through Syria and Turkey, and down into today's Iraq and the Zagros flanks in Iran, toward the Persian Gulf. Most of it was fertile and well-watered compared to neighboring mountains or the Arabian desert (Figure 13.1).

Teaching a university class today, I lecture that becoming agrarian – that transition from Paleolithic to Neolithic – was one of the two most significant transformations in human history. (The other was the industrialization of the last three centuries.) But much I learned as a student is no longer taught because our understanding of this period has changed and no doubt will continue to change with new discoveries, especially in regions not as well studied as the Middle East. The origin of agriculture in the Middle East is still dated to 10,000 years ago, or a little earlier, its oldest-known sites still in the Fertile Crescent. But that is no longer regarded as the sole origination point for agrarian life. There is good evidence of independent agricultural beginnings in central China, the New Guinea highlands, Mesoamerica, the central Andes, the Mississippi basin, western Africa, and southern India, occurring at various times between about 10,000 and 4,000 years ago (Bellwood 2005). Agriculture did not begin once but many times. Furthermore, it now appears that sedentary community life preceded agriculture, at least in the Fertile Crescent.

By 14,000 years ago, millennia before agriculture, there were villages in the Levant, to all appearances intended for year-round occupation. The inhabitants were communities of hunters and gatherers called by archeologists the Natufian people (Bar-Yosef 1998). Their dwellings were not haphazardly placed but aligned along hillslopes, well built with floors below ground level and stone walls to support roofs of brush or hide. Each village had a cemetery holding richly decorated bodies, some

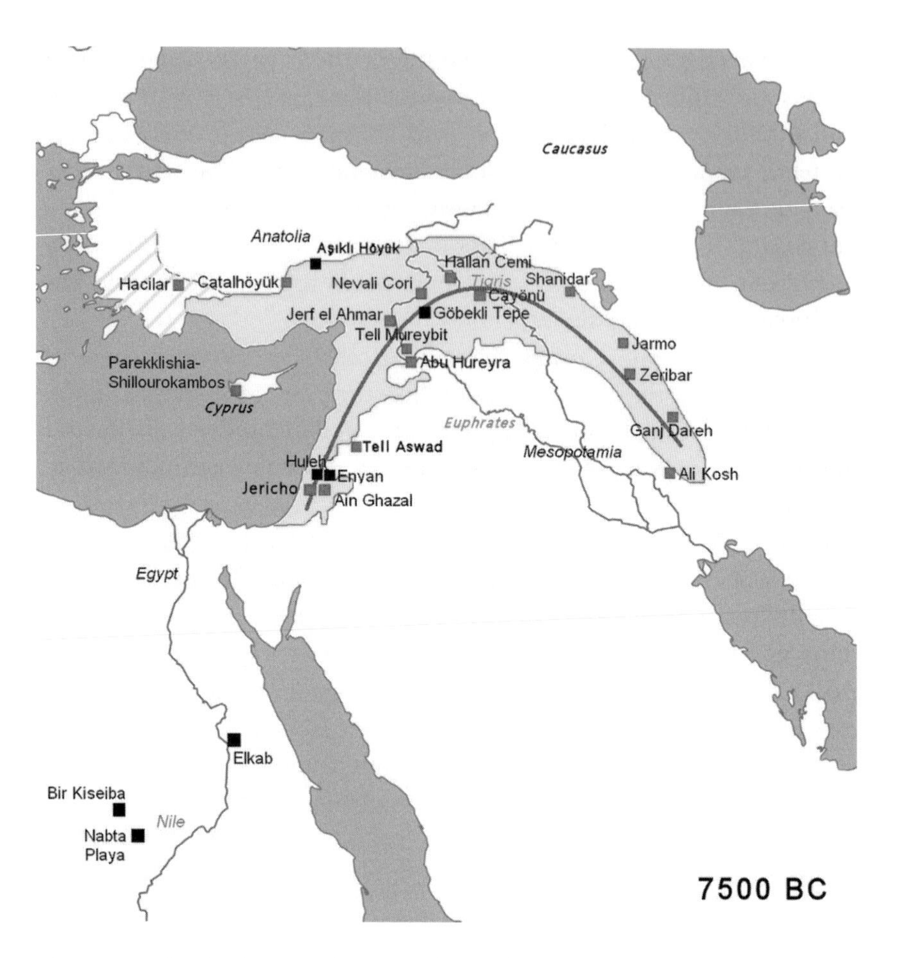

Figure 13.1 Fertile Crescent circa 7500 BC, including archeological sites. At this time, Mesopotamia was not yet an area of habitation by farmers.
Source: Creative Commons Attribution-Share Alike, GNU Free Documentation License

graves containing dogs. Located in oak woodland with a felicitous climate, there were ample wild plants for foraging. Villagers had stone mortars and pestles to grind acorns and almonds into paste. They used sickles, bone handled with flint microliths, as blades to cut edible plant stalks including wild wheat, which was useful for straw, as well as grain that was ground into flour and baked as flatbread. Hunters with bows and arrows preyed on gazelles and other game, including pheasants (Mithen 2003). There was no need to grow food because it was abundant nearby. In the Fertile Crescent, agriculture and sedentary life were not like chicken and egg. Settlements came first.

This is a repeated pattern. Hunter-gatherers living in productive regions and emphasizing plant or fish-rich diets are often sedentary or semi-sedentary. The Jomon people of Japan are a good example, hunters and gatherers living in settled communities as early as 11,000 years ago, remaining entirely reliant on wild food from plentiful forest and coastal resources, also makers of the earliest-known pottery. Not until about 2,500 years ago were they replaced by migrants from China or Korea bringing rice cultivation. The California Indians are another example, as are the builders of elaborate timber-framed houses in fishing communities in today's Washington State and British Columbia.

Improving climate had produced an abundance of plants and animal prey, making pre-agrarian villages possible, even attractive. Population has usually been thought to have increased as a natural consequence of settled life since it was no longer necessary to limit the number of babies that must be carried from site to site, however that has recently become a point of controversy. Estimates of prehistoric population growth, based on statistical analysis of the radiocarbon record, show that transitioning farming societies experienced the same rate of growth as contemporaneous foraging societies (Zahid, Robinson, and Kelly 2016). However that is resolved, the abundance of wild plant food was interrupted by the Younger Dryas, leading some Natufian people to revert to their mobile lives until the warmth resumed. Jericho, known from the Bible, was first settled when local food was returning in quantity. Its earliest inhabitants included Natufian people, moving back toward settled life.

Jericho

The present-day Palestinian city of Jericho, near an oasis in the Jordan Valley, is walking distance from Tell es-Sultan, a large mound formed by the accumulated ruins of successive occupation levels of ancient Jericho. The tell was first excavated in the nineteenth century in a failed search for the city's fabled defensive walls, said to be destroyed during the Israelite invasion of Canaan under Joshua. More thorough and careful excavation in the twentieth century, especially by the British archeologist Kathleen Kenyon (1906–1978), revealed at least 20 successive settlements. The earliest occupation level, with Natufian flint tools and the remains of a small building, dates back 11,000 years, shortly after the Younger Dryas had ended. Probably these Natufian hunter-gathers were temporary residents, but as wild vegetation again became abundant and gazelles plentiful, occupation became year-round. Through successive periods of growth, decay, abandonment, and rebuilding, Jericho may be the oldest settlement persistently occupied into modern times, spanning the transitions from the Paleolithic to Neolithic to the Bronze and Iron ages.

At a higher (and later) level, dated to about 9000 BC, are the remains of round houses constructed from sun-dried mud brick. The dead were buried under the floors of buildings. Human skulls are found with facial features reconstructed in plaster and eyes set with shells, perhaps indicating a cult practice. Structures covered an area equivalent to one square city block, with a stone wall over 3.6 meters (12 ft) high and 1.8 meters (6 ft) wide at the base. There was a conical stone tower 8.5 meters (28 ft) high and 9 meters (30 ft) in diameter at the base, with an inner staircase leading to the top of the wall (Figure 13.2). It is one of the earliest monumental structures known. The wall and tower are

Figure 13.2 Stone tower at Tell es-Sultan, Jericho, 9 meters in diameter, constructed ca.11,000 years ago. It is one of the oldest examples of monumental architecture. Photograph by Avi Deror, GNU Free Documentation License, Wiki Creative Commons

thought by some to be defensive structures, but others think the wall protected the site from flooding, which became unnecessary as the tell grew higher over the centuries. The tower may have had some ritual or astronomical use. The town's inhabitants had no domesticated plants or animals, perhaps cultivating wild ones. Meat was still supplied by hunting wild gazelle.

The history of Jericho's settlement is punctuated by periods of growth and abandonment. There is a town dated about 6000 BC, now based on irrigation farming. This population had domesticated emmer wheat, barley, and pulses (beans, peas). Meat was still supplied by hunting wild gazelles though possibly there were domesticated sheep.

Pottery appeared in Jericho less than 7,000 years ago, and bronze about 2,000 years later. By 4,000 years ago, the city was again prosperous, surrounded by a high sloped wall of beaten earth supported at its base by a stone retaining wall 20 feet high. Many tombs found outside the city contain alabaster and bronze, scarabs and jewelry, wooden objects, reed mats, and baskets. The city was destroyed again, but it was inhabited in the thirteenth century BC, when Joshua reputedly led the Israelite invasion of Canaan, however there are no signs of destroyed walls from that period (Avigad 1974).

New discoveries in the Fertile Crescent have broadened our understanding of the profound cultural changes seen at Jericho. Göbekli Tepe, nearly as old as Jericho, was first located in the 1960s but not seriously excavated until the 1990s. Here are massive rectangular pillars carved out of limestone, some eight feet high and weighing seven tons, dragged from quarries up to 100 meters away and sunk into the hill to form circles. Many of the pillars are carved with wild animals. Five circles have been excavated but 15 more are indicated by ground-penetrating radar. At the center of each circle are two larger pillars. This was constructed before the domestication of food; the workers ate wild plants and animals. With no trace of dwellings places, Göbekli Tepe is assumed to have been a ceremonial center rather than a town.

Such pre-agrarian sites were transformed into, or replaced by, settlements in which the inhabitants planted crops and husbanded animals. Çatalhöyűk in southern Turkey is one of the most impressive sites of the early Neolithic. Discovered in 1958, James Mellaart began evacuating during the 1960s, followed by Ian Hodder. Their finds include painted walls, bulls' heads modeled in clay, human burials, and figurines as well as mirrors made from obsidian, and daggers with carved bone handles. Clearly an agricultural village with population estimated in the thousands, it flourished about 9,000 years ago but was abandoned before the Bronze Age.

Architecturally, Çatalhöyűk is unlike any other site. People lived in contiguous rooms with plastered walls but no doors or windows. These homes, sharing common walls, had no streets between them (Figure 13.3). Room entry was through an opening in the ceiling, reached by a ladder, so travel from one room or home to another was across the continuous rooftop of the community. This seems to have been an egalitarian society, at least neither the structures nor artifacts suggest social stratification. Dead people, their heads removed, were buried within the village. Wheat, peas, and barley were cultivated; almonds, pistachios, and fruit collected from trees in the surrounding hills. Sheep were domesticated and possibly cattle, but hunting continued to be a source of meat. The town was one node of a network trading obsidian, flint, and Mediterranean seashells (Hodder 2011).

Figure 13.3 Contiguous rooms at Çatalhöyük, after the first excavations. Photographed by Omar Hoftun, Creative Commons Attribution-Share Alike 3.0 Unported license

Why Agriculture?

In the mid-twentieth century, when archeologists thought that agriculture had a unique origin in the Fertile Crescent ten millennia ago, there was little constraint on speculation about why this happened then and there. Perhaps one man, a prehistoric Einstein, invented it. Given the impressive record of invention during the Upper Paleolithic, this was not farfetched.

The foraging life had sustained the genus *Homo* for 2.5 million years, so why the sudden change? Early on, it was presumed that agriculture, once available, was intrinsically preferred to hunting and gathering. But as ethnographers began studying today's remaining hunting-gathering societies, they found the foraging lifestyle felicitous for those still practicing it. Much of their day is free for leisure, they are politically egalitarian, living in balance with nature, content with few material possessions. The option of farming and herding can be grueling and boring work without let-up, living in hierarchical societies where a few are well off and most not. There was more disease, sometimes derived from domesticated animals, also from the small pests, so-called commensals like fleas, lice, rats, and mice that infest permanent settlements, and more contagion in densely populated agrarian societies. Farmers' bodies may be physically stunted compared to foragers living off the land. Skeptics began (Sahlins 1968) and continue (Scott 2017) asking, why would anyone choose backbreaking agriculture over the free wandering of a forager? Of course, one can slant the question differently: Why would anyone choose the dangerous or arduous life of a hunter-gatherer over a safe, regular routine in the fields near home, or the leisurely days of a shepherd? Either way, the question may be misplaced since individuals do not choose into what circumstances they are born. You take what you get and adapt to it.

It is easier to see advantage from agrarianism at the group level. Cultivation provides far more calories of food per acre of land than foraging, allowing population to expand. We cannot say if Neolithic people sought this goal or not, but it happened. Population density increased during the agrarian transformation.

Even at the group level, we should not frame this as an either/or choice. Many groups mixed foraging with agriculture, pursuing them simultaneously, or at various times moving between one mode and the other. Hunters and gatherers did modify their environments, notably burning brush to encourage new growth of desirable plants or animals, scattering some seeds to be gathered at the next visit. Among the diverse Indian tribes of California, for example, were hunters and gatherers who, at the same time, managed their landscape to enhance its edible resources by burning, pruning, sowing, weeding, digging, thinning, and selective harvesting (Anderson 2005). In what is today Yosemite National Park, the Miwuk and Paiute Indians tended black oak and pinyon pine groves, removing dead or diseased branches. In fall, Yosemite Valley was burned to get rid of competitive species, especially unwanted conifers (Liz Williams, personal communication).

Once agriculture became established, humans embarked on ten millennia of artificial selection of plants and animals. Farmers saved for replanting those seeds that were plumper or otherwise most desirable. Archeologists can distinguish wild from domestic seeds, a common example being in ears of grain. Wild ears are brittle, spontaneously shattering when ripe so their seeds scatter to the ground. A small portion of wild ears lacks the shattering trait. These non-shattering ears became the common form of domestic wheat, requiring threshing to reach their grain. This came about inadvertently. When humans began cutting grain stalks, the earliest shattering ears had already dispersed their seeds, so harvesters oversampled ears that lacked the shatter trait. After many harvests, the non-shattering variant became the common form. Without human management the domesticated forms are unable to reseed themselves. (See Diamond 1997 for an extended discussion of artificial selection).

Obviously, agriculture required a warmer climate than the Ice Age, but this was not much included in early explanations of its origin because so little was known about prehistoric climate. Research on past climates accumulated rapidly after the 1980s, pursuant to public concern about anthropogenic global warming, bringing its importance to the fore.

By 15,000 years ago, the great northern ice sheets were receding as Earth's climate warmed and became wetter. Atmospheric CO_2 rose, conducive to plant growth. These trends were interrupted by the Younger Dryas, 12,900 to 11,700 years ago, returning to severe glacial conditions for over a millennium, but afterward temperature resumed warming and has remained at the "comfortable" and unusually stable level that we have known in historic times (Figure 13.4). Human diets, which in glacial conditions necessarily depended heavily on meat, were increasingly supplemented with plant food.

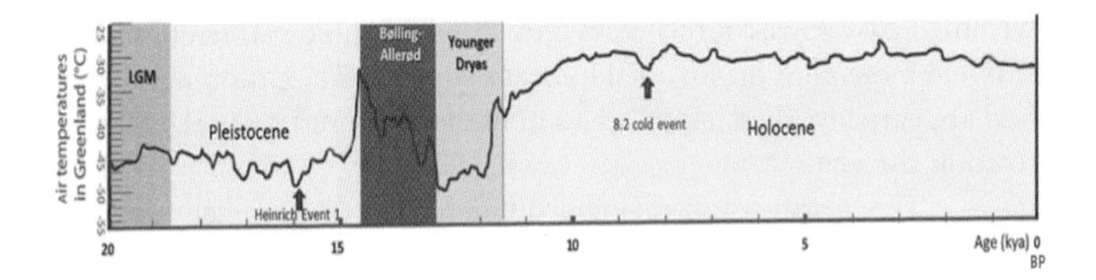

Figure 13.4 Air temperature over the past 20 thousand years, as estimated from Greenland ice core chronology (Rasmussen et al. 2006), showing the transition from the Pleistocene to today's Holocene epoch.
Source: Wikipedia Creative Commons Attribution-Share Alike 4.0 International license

Some archeologists proposed that the cold Younger Dryas diminished the natural food supply on which sedentary Natufians depended, spurring them to plant founder seeds, thus beginning domestic agriculture (Bar-Joseph 1998). This proposal has since been contested with evidence that there was no food insufficiency during the Younger Dryas, and besides, purposive cultivation did not begin until the climate was again warming (Balter 2010). Within another two and a half millennia, starting around 6500–5500 BC, farming (and pottery) spread from the Fertile Crescent into Europe and South Asia, and below the Mediterranean into Egypt and North Africa (Bellwood 2005).

Other Centers of Origin

By the late twentieth century, it seemed implausible that the Fertile Crescent was the sole font of agricultural practices in far flung, disconnected sites. These were increasingly discovered to have their own long histories of raising domesticated plants and animals whose wild forms were not found in the Middle East but were indigenous to their own regions. China, for example, had wild millet in its north and wild rice in the south, both converted to recognizably domestic forms not very long after farming began in the Levant, possibly earlier. (Chinese dates are insecure because of degradation of crop remains in its wetter climate and, until recently, far less archeological work than in the Middle East.) The earliest known domestic millet, as well as grinding stones, stretching across north China down to the Yellow River, is dated 8,000 years ago. The earliest rice paddy cultivation has been found in the middle and especially lower Yangtze River region, along the coastal wetlands, as early as 7,700 years ago (Zong et al. 2007). Strengthening summer monsoons during the Holocene supported rice farming. Pottery had appeared by then, earlier than in the Fertile Crescent, perhaps used to cook the rice.

This north-south regional difference between millet- and rice-cultivating regions corresponds to a population difference. Genome-wide data from 26 ancient individuals spanning 9,500 to 300 years ago indicate that in the past, genetic differentiation from north to south was greater than at present. There was a major episode of mixture involving northern Chinese ancestry spreading south after the Neolithic, erasing the earlier differentiation (Yang et al. 2020).

China had soybeans, pigs, dogs, and chickens that originally came from mainland Southeast Asia or the Philippines. As in the Fertile Crescent, all these wild foods were collected and used before their supplies were artificially increased through cultivation. On the other hand, wheat, barley, sheep, goats, and horses were domesticated elsewhere before their adoption in China (Liu and Chen 2012).

Shortly after Europeans became aware of the Americas, the Spanish conquistador Francisco Pizarro arrived in 1533 at Cuzco, the 11,200-foot (3,400 m) high Andean capital of the Inca empire. He learned that the Incas raised guinea pigs, ducks, llama, alpacas, and dogs, cultivating corn, squash, potatoes, tomatoes, peppers, and cotton (Ayala 1995). These domesticates far predated the Inca empire. Squash was first planted 8,000–9,000 years ago, peanuts and quinoa about 8,000 years ago, and cotton 5,500 years ago. Stone hoes, ground stone bowls and pallets, furrowed garden plots, and small-scale irrigation canals have been found from the same period (Dillehay et al. 2007).

Mesoamerica, the region from central Mexico through Central America, is recognized for its important pre-Columbian societies. No doubt the Aztecs are most famous of these, but given their late and short reign (1300–1521 AD), they are unimportant here. Farming began long before, as early at 7000 BC, eventually domesticating maize, cacao, beans, tomato, avocado, vanilla, squash, and chili, as well as the turkey and the dog. Most of these plants were not found in the Andes, though maize reached the area early. There is little evidence of early contact with Peru, which in any case would have been difficult, so Mesoamerica itself appears as an independent point of early agricultural origin. The combination of maize, beans, and squash – the "Three Sisters" – has become a staple diet among indigenous Americans beyond Mesoamerica. Today, maize (corn) is one of the most widely distributed of the world's food crops, though most is not consumed by humans, instead fed to livestock, converted into biofuel, or used as raw material in industry (Pollan 2006).

Kuk Swamp is the site in highland New Guinea with best evidence for early and independent cultivation at least 6,500 years ago, likely earlier. The oldest archeological features are narrow channels, pits, and postholes, features consistent with planting, digging, and tethering of plants and localized irrigation and drainage of a cultivated plot. Microfossils of taro and banana were found, perhaps the most important food staples in the Highlands before the introduction of the sweet potato after the much later European exploration of the Pacific (Denham et al. 2003).

During the Last Glacial Maximum when sea levels were low, New Guinea and Australia were connected as the single landmass of Sahul, but as ice melted and the sea rose, they became separated by the newly formed Torres Strait. It is peculiar that New Guineans became farmers, while nearby Aborigines of northern Australia remained hunters and gatherers, not for lack of knowledge because the groups likely knew of each other. Asking why the Australians, too, did not become farmers, Peter White of Sydney University suggests that they were "simply too well off to bother about agriculture" (quoted in Mithen 2013: 338). Another reason may be the extreme aridity and infertile soil over much of that flat continent.

When the British occupied Australia and introduced their own agricultural practices, in the process subjugating and largely eliminating the indigenous population, they introduced a nationalist mythology emphasizing the primitive quality of the native society being displaced, one that had not yet mastered the elements of farming and land management. (It was an antipodal version of the European account of "primitive" Native Americans.) A recent bestselling book, *Dark Emu: Aboriginal Australia and the Birth of Agriculture* (Pascoe 2018), has stirred considerable self-reflection among European-derived Australians, presenting as it does documentation by early settlers of existing systems they found on arrival of food production and land management. While details have been challenged, the dubious nature of the mythology is compelling.

Explaining the Transformation

For explanatory purposes, it was crucial to discover that the Fertile Crescent was *not* the sole seedbed of agriculture, that other locations around the world began cultivating independently between 10,000 and 4,000 years ago. No longer did it make sense to seek a unique cause in the Fertile Crescent. Causes lay in conditions common to those multiple, nearly simultaneous loci. Not surprisingly, they all had abundant water, at least seasonally. The Holocene's warmer and stable climate was newly exploitable by people with modern cognitive and language ability, and food processing skills honed during the Upper Paleolithic. A proviso was having access to indigenous plants and animals adaptable to domestication. This was the deficit that prevented Egypt from originating its own agriculture, despite suitable people and the Nile's ideal flood regime. There were no locally adaptable plants or animals. Not until about 7,500 years ago, after Greece and Italy, did Egypt become agricultural when a full agrarian economy with pottery was introduced from the Levant. Apparently, the Sinai desert had been an impermeable barrier to earlier diffusion (Bellwood 2005).

Cultivation was later to develop in the New World than the Old if only because people arrived far later on the North American continent, and the maize cob took many millennia to be domesticated to a useful size. If that were not sufficient reason, the lag in New World farming may also owe something to its relative scarcity of wild plants and animals suitable for domestication (Diamond 1997).

With the recognition of multiple origins and the importance of climate, Peter Richerson and his colleagues have advanced the provocative proposition that agriculture was impossible during the Pleistocene but nearly inevitable during the Holocene (Richerson, Boyd, and Bettinger 2001; Richerson, Gavrilets, and de Waal 2021). One might picture the conversion very much on autopilot, though the authors do not use that term. Whether or not that is true, the question of agriculture's origin seems to have changed from *How did it happen?* to *Why wouldn't it have happened?* Perhaps like the Aborigines of Australia, some hunters and gatherers simply were not interested.

Once established, agrarian practices diffused through other regions, some quite far away. By 6,000 years ago, farming from the Middle East had spread as a major form of subsistence from the Atlantic coast to the steppes of Central Asia. Was it the *concept* of agriculture that had spread, adopted as a new lifestyle by indigenous hunters and gatherers? Apparently not. Modern Europeans are not genetically descended from Europe's diverse hunters and gatherers but do carry DNA from Near Eastern migrants bringing their agriculture. The people of today in Europe, the Middle East, and much of central Asia are genetically similar to each other, which was not the case 10,000 years ago. This homogeneity may be a consequence of farmers from Anatolia moving westward, replacing the prior residents, perhaps because of their higher rate of reproduction (Reich 2018).

There is another relatively homogeneous group in East Asia, their DNA recognizably different from that of West Eurasians, the genetic boundary lying in central Asia (Reich 2018). This may reflect the independent diffusion of agriculture in East Asia. Possibly all centers of agricultural origin nucleated genetically homogenous regions around them.

Barriers to Diffusion

Jared Diamond stresses the importance of a continent's directional axis in the diffusion of domesticates, with Eurasia favored by its long east-west extent. Locations at the same latitude share the same-day length and its seasonal variations. To a lesser degree, they have similar diseases, temperatures, precipitation, and types of vegetation. Portugal, northern Iran, and Japan, far distant but at the same latitude, are more similar to each other in climate than each is to areas lying 1,000 miles (about the distance from Florida to New York City) due south. On all continents containing tropical rain forests, they occur within 10 degrees of the equator. Mediterranean-like scrub habitats lie between 30 and 40 degrees of latitude. Since the germination, growth, and disease resistance of plants are adapted to these features, they more easily spread along parallels of latitude than meridians of longitude (Diamond 1997).

Perhaps more important are physical features of Earth's surface that block diffusion, most obviously oceans but also land barriers. China is effectively boxed in by the Himalayas, the Tibetan Plateau, the Gobi Desert, and the Mongolian steppes, all of which separate East Asia from western and central Eurasia. These can be crossed but with difficulty. Eventually the Silk Road traversed the Gobi, invaders entered from Mongolia, and Buddhism reached China from India, perhaps carried through Himalayan passes. Far earlier, Peking Man had immigrated from Africa to China, almost certainly along coastal routes. But despite such "leaks," these physical barriers effectively separated the East Asian population from people to the west, forming separate gene and culture pools.

There was unobstructed travel from the Atlantic to India and into North Africa until blocked by the Sahara Desert. (The Sahara was not always so arid, having lakes and being home to game between 11,000 and 6,000 years ago.) The Mediterranean Sea looks on a map like a barrier, but more often was a convenient water route between points on its shores, as when it would eventually be encircled by the Roman Empire.

The concept of racism was based on skin color and facial features that were recognizably different among typical inhabitants of East Asia, Western Eurasia plus North Africa, and sub-Saharan Africa. This lends itself to extreme stereotyping, ignoring considerable physiognomic variation within each "race." There is more genetic variation among African peoples than among non-Africans (Reich 2018). Still, it seems clear that separate genetic/cultural zones developed within geological barriers, and there was limited diffusion between these zones. The distribution of language families shows similar effects (Bellwood 2005).

Sir William Jones (1746–1794) was an English lawyer and scholar of languages. While serving as a judge in Bengal he became proficient in Sanskrit, the primary liturgical language of Hinduism, once the language of ancient India. Noticing common verb roots and forms of grammar in Sanskrit, Greek and Latin, and perhaps Gothic, the Celtic languages, and Persian, he suggested that they had a common source, which came to be known as Proto-Indo-European. Jones was not first to recognize similarities, but his hypothesis became widely known and the basis for research (Auroux 2000).

It is now accepted that most of the languages of Europe and through to India are derived from this source language, itself no longer surviving. The major interruption in this spread of Indo-European is a zone of Semitic languages (part of the Afro-Asiatic family), including Arabic and Hebrew, in the Middle East and North Africa. East of the great Tibet-Gobi-steppes barrier is an independent language family, Sino-Tibetan, mostly Chinese, and there are also central Asian Altaic languages.

After writing was invented in China about 4,000 years ago, it diffused throughout the Eastern Zone, being incorporated into Japanese, which is not a Sino-Tibetan language, producing the strange situation today where people from China and Japan can read some of each other's writing but cannot speak each other's language. The modern alphabet of Semitic-Greek origin appeared about 3,000 years ago, spreading to virtually all cultures of the Western Zone, although remaining quite distinct from writing in the Eastern Zone.

Long term importance of this Tibet-Gobi-steppes barrier may be illustrated by jumping ahead in our history. Alexander the Great (356–323 BC), by the age of 30, had marched his army from Macedon to Northwest India without encountering unsurmountable military or physical blockages, conquering the full extent of the "known world." As far as he knew, there was nothing interesting farther to the east.

In the four centuries around the beginning of the Christian era, the Roman Empire and China's Han Dynasty were the world's two major hegemons. On opposite sides of Eurasia, separated by the Tibet-Gobi-steppes barrier, they were barely if at all aware of one another's existence before the first century AD.

What's Next?

After the Ice Age, the nearly complete adoption of agrarian life within about five millennia became the substrate upon which towns grew into cities, into kingdoms, into empires (Scarre 2018). These are the social entities of what we call history, the written accounts of our past, occurring since three or four thousand years ago. Their great dramas are what most people think of as the past, though occupying a sliver of human time. How did civilizations of the historical era arise from settlements devoted to cropping and herding?

14 RISE OF CIVILIZATIONS

Elizabethan playwright Christopher Marlowe wondered about Helen of Troy, "Was this the face that launch'd a thousand ships?" Homer's *Iliad* implicates her as the *casus belli* of the Trojan War. The story goes that a peace delegation of Trojans was visiting the Greek city of Sparta, where Menelaus was king and the beautiful Helen his wife. Prince Paris of Troy was smitten and either abducted her or she eloped with him. Enraged Menelaus, in coalition with his brother King Agamemnon of Mycenae, and other Greek cities, assembled a great fleet to retrieve his wife, marking the start of the ten-year siege of Troy (*Ilium*) in Asia Minor.

Homer's epic, originally an oral narration, perhaps written down in the eighth century BC, has many heroes, none greater than Achilles, who slays Troy's champion Hector and shamelessly ties the fallen body to his chariot, dragging it outside the city's high walls in triumph. Still Troy holds out. Finally, under Odysseus's stealthy plan, the Greeks feign a return home, leaving behind a giant wooden horse with soldiers secreted inside. Jubilant Trojans, thinking the siege over, drag the horse into their city. That night the hidden Greeks open the city gates for their comrades to enter, slaughter the Trojans and destroy their city.

Readers have since debated whether Helen or the various heroes, for that matter Homer, were real people. Had there actually been a Trojan War so long ago? Was Troy a real city, and if so where was it? Alexander the Great and Julius Caesar each visited a site reputed to be Troy.

No early archeologist was more avid a believer in the historicity of Homer than a German American businessman and speculator named Heinrich Schliemann (1822–1890). Wealthy enough to retire at age thirty-six, he then dedicated himself to finding Troy. Taking the advice of an English expatriate, Frank Calvert, they began collaborating in 1871, digging at Hissarlik in Turkey, which showed several layers of habitation. It was careless excavation. Thinking that ancient Troy must be at the lowest level, Schliemann had his workers dig hastily, destroying some of the middle layers to reach the deepest remains, where they found fortifications (Allen 1999).

By this time, Schliemann and Calvert had fallen out, and the divorced Schliemann found himself a new wife, Sophia Engastromenos, thirty years his junior. He often portrayed her as an active partner in the dig, but she was not really avid, and Schliemann was not a consistently truthful source. Anyway, when a cache of gold and other objects appeared in 1873, Schliemann adorned Sophia with them, declaring this was the treasure of King Priam of Troy (Figure 14.1). Perhaps because of the publicity, the Turkish government revoked Schliemann's permission

Figure 14.1 Sophia Schliemann wearing treasures discovered at Hissarlik.
Source: Wikipedia

Figure 14.2 "Mask of Agamemnon," made of gold, discovered by Heinrich Schliemann in 1876 at Mycenae, now at the National Archaeological Museum of Athens. Attribution: Die Buche, Wikipedia Creative Commons

to dig and sued him for a share of the gold. Schliemann smuggled the treasure out of the country and began digging at Mycenae. He found gold there too, most famously what he called the "Mask of Agamemnon" (Figure 14.2).

Later excavation at Troy showed the layer he had identified as Homer's city was in fact a thousand years too old. If the *Iliad's* Troy was really there, it would have been at one of the higher layers that Schliemann had plowed through and destroyed. The Mask of Agamemnon is also too old to have been worn by Agamemnon of the *Iliad*, judging by the layer where it was found. In view of Schliemann's poor record of truthfulness, there is some question of its authenticity (Gere 2011).

Leaving aside whether Schliemann was truly at the site of Troy, there is no doubt that he was digging through settlements more sophisticated than the early agricultural villages of the last chapter, as was Homer's storied Troy. Having reached a higher level of complexity and achievement, archeologists thought of these sites as ancient *civilizations*, and these became the focus of their discipline.

Archeology's Classic Civilizations

The scholarly discipline of archeology began around the seventeenth century with antiquarian collectors of old art and artifacts, often displayed in cabinets of curiosities. It was stimulated by excavations at Herculaneum and Pompeii, Roman cities buried by lava and pumice from the eruption of Mount Vesuvius in 79 AD, preserving them nearly intact. Essentially ignored over the centuries, digging at Herculaneum began in the early 1700s, the main purpose to find valuable marble statues that had been buried by the falling ash. As patrons became aware of the stunningly detailed preservation of these places, digging was extended to Pompeii, the looting motive ceding place to historical curiosity. The site became and continues today to be a major tourist attraction. Vesuvius's eruption, included hot gas, so quickly hit Pompeii that street and home scenes were revealed as they had been at the moment of catastrophe, showing intact mosaics, murals, graffiti and figurines, some so pornographic that they were closeted from public view. Food was found on tables. Corpses encased in pumice hardened by rain were caught at the moment of death.

Another stimulus came from Napoleon's 1798 campaign in Egypt, which included a large corps of scholars to study the art and history of the country, inspiring a wave of Egyptomania in Europe (Wilkinson 2020). French soldiers found a granite stele, the Rosetta Stone, which soon was captured by the British and brought to London. It was inscribed with three versions of a decree issued in 196 BC, one written in ancient Egyptian using hieroglyphic characters, another in ancient Egyptian Demotic script, and a third in ancient Greek. Many European scholars could make out the Greek, so this was obviously a clue to deciphering hieroglyphics, beginning a race to do so. The winner was Jean-Francois Champollion (1790–1832), who announced his translation of the Egyptian scripts in 1822, thus opening to view the abundant hieroglyphic inscriptions of the ancient Nile civilization. Key was Champollion's recognition that the royal name, Ptolemy appeared repeatedly in the Greek version. Locating these repetitions in the other versions was the opening step in their translations (Buchwald and Josefowicz 2020; Wilkinson 2020).

While Enlightenment scholarship shaped the scientific core of archeology, perhaps as important was the Romantic Movement and its glorification of the Classical era, its ruins and heroes. There is no better example than Schliemann's search for Troy. The rise of Nationalism was also important as states sought to demonstrate their roots or their colonial achievements by displaying ancient finds in national museums such as the Elgin Marbles in London's British Museum. (Removed in 1803 from the Parthenon in Athens by Thomas Bruce, 7th Earl of Elgin, they have not been returned to Greece despite repeated requests.)

The nineteenth century was a heyday for European archeologists focused on the classical civilizations of the Middle East with major projects in Egypt, the palace of Assyrian ruler Sargon II, the ruins of Babylon and Nimrud, and as we have seen, Schliemann's search for Troy. Digging was haphazard, intent on finding impressive artifacts. Among the improvements by the end of the century was the use of stratigraphy, taken from paleontology, to place artifacts and settlement layers in chronological order. Emphasis shifted from finding valuable objects to reconstructing the history of a site, and the realization that all artifacts, even broken pottery, were important for this end (Silverberg 1997).

It was a field for affluent amateurs. Leonard Woolley (1880–1960) wrote in his autobiography, "I had never studied archaeological methods even from books . . . and I had not any idea how to make a survey or a ground-plan" (1955: 15). Nonetheless, he became one of the first to work in a methodological way and was among the most successful archeologists of his generation. In 1912–1914, with T. E. Lawrence (shortly to become "Lawrence of Arabia"), he excavated the Hittite city of Carchemish, at the same time the two of them were spying on German railroad construction for British Intelligence. After the Great War, Woolley led a British American expedition to Ur in Mesopotamia, uncovering some of the ancient city's splendid accomplishments, especially grave goods at the royal cemetery, including the Bull Headed Lyre, dated about 2500 BC, one of the oldest known stringed instruments. One of Woolley's assistants, Max Mallowan (1904–1978), met Agatha Christie (1890–1976) at the dig. She was recovering from a marriage that had foundered on infidelity. They were wed shortly afterward, and she regularly went with Mallowan on his own excavations in Egypt. Her detective novel, *Murder in Mesopotamia* (1936), one of sixty-six she would eventually publish, was inspired by discovery of the royal tombs at Ur. Woolley may have been the first to suggest that the Bible's Flood story was adopted from Mesopotamian mythology.

Probably Howard Carter (1874–1938) is one of the most famous names in Egyptology (apart from Agatha Christie). While the great pyramids were obviously of initial interest, all of them had been robbed as had nearly all other important tombs in Egypt. New Kingdom pharaohs, realizing the vulnerability of ostentatious burial places, began hiding them, dug into cliff walls in the Valley of the Kings. Carter worked there intermittently for some years, finding tombs there had also been pilfered. In 1922, toward the end of his stay, his water boy literally stumbled on stones at the top of a hidden flight of stairs leading to a still sealed tomb. It turned out to be the burial place of an unimportant pharaoh named Tutankhamun (ca. 1342–1325 BC), who died as a teen, apparently creating the need to hurriedly arrange interment in a minor tomb. Still, when Carter, accompanied by his sponsor Lord Carnarvon, first peeked through the sealed door, and Carnarvon asked if he could see anything, Carter famously replied, "Yes, wonderful things!" Thousands of these objects, including King Tut's mummy, will be permanently housed at Cairo's new Grand Egyptian Museum. Carter became famous, and Tut's golden burial mask is recognized around the world (Figure 14.3).

This is the briefest tour of famous civilizations explored as archeology was getting its legs, mostly in the Middle East, which remains the best explored region, but we depend as well on work farther east in Asia, in Africa, Australia, and the Americas.

What Exactly Is a Civilization?

There is no consensually accepted definition of civilization (Trigger 2003). Roughly speaking, the term refers to an advanced agrarian society that has built cites with monumental architecture, defensive walls, advanced arts and crafts, planned religious and ceremonial centers, accomplished metallurgy and ceramics, animals not only to eat but as beasts of burden, calendars, astronomical observation, writing, mathematics, and intensive irrigation projects. Formal hierarchy is present, often in the form of hereditary kingship, with a small elite, often warriors and landowners, and a large peasantry, reflected in differences in housing, personal adornment, armaments, and burials. There are armies and warfare, laws and taxation, conscription and slavery. This is not an exact list of requirements to qualify as a civilization, and not every early civilization had all of them, but the clustering is fairly reliable.

Figure 14.3 Funerary mask of Tutankhamun, colored gold and blue, at the Egyptian Museum in Cairo.
Source: Wikipedia Commons, GNU Free Documentation License, Version 1.2

Civilization is often contrasted with barbarism, the latter an invidious label from the perspective of civilized historians, though likely not from that of the barbarians themselves. No doubt monumental cities were impressive to see but oppressive and unhealthy as well, and neither intrinsically superior, more desirable, nor even more sophisticated (excepted in a particular sense) than the tribal cultures of those not associated with them. It is like contrasting the appeal of agriculture versus hunting/gathering as lifestyles, which may be more a matter of taste than intrinsic quality. Even during the era of great civilizations, some of the most powerful social forces were barbarian, one example the Huns of Central Europe under Attila (408–453), whose horsemen seriously threatened the Eastern and Western Roman empires. Western Rome received its death blow from the army of Odoacer, a soldier of barbarian background. The Mongols are another example, their horsemen under Genghis Khan (1162–1227) conquering most of Eurasia, his successors including Kublai Khan becoming the opulent Mongol Dynasty of China.

Mesopotamia

The earliest known civilization is Sumer, a cluster of cities in southern Mesopotamia, the "land between the rivers," a flat, nearly featureless floodplain of the Tigris and Euphrates in modern Iraq. Today arid, in antiquity much of it was muddy with few plants, subject to unpredictable and severe flash flooding from the mountains. It was not inhabited when agriculture began in the Fertile Crescent, but some five millennia later the land near the head of the Persian Gulf was being cultivated, and by 3500 BC it held several city states, most importantly Uruk and Ur. There were not many natural resources. Apart from water and fertile land there were mud, clay, and reeds, but these were put to good use. Mats and baskets were made from reeds. Clay was used for pottery. Bricks made from mud, hardened by drying in the sun, were the construction material for all structures in the area, most impressively city walls and especially temples in the form of ziggurats, which emulated mountains, supposed home of the gods (Figure 14.4).

Cuneiform, one of the earliest forms of writing, was inscribed on wet clay tablets with a reed stylus producing wedge-shapes marks, which could be erased by rubbing with water or dried and baked for permanency. Like all early writing, it is based on the rebus concept of repurposing elemental pictures of objects into symbols of the spoken names of those objects. Its many symbols and conventions to represent words or syllables required the intensive and uniform training of a class of scribes. Tablets have been recovered in abundance and deciphered. Originally cuneiform was used for tabulating inventories and tax records, later for chronologies and eventually narratives; its 90°-rotated characters evolved to be used for other Mesopotamian languages (Figure 14.5). Another literary device was the small cylindrical seal, inscribed with individualized designs and pictures, a sort of signature for important people, which could be rolled across a wet clay surface. It is unclear where the Sumerians came from as their language does not resemble any others that are known.

The *Epic of Gilgamesh*, dating from the second millennium BC, was first written in the Sumerian language but is known from several surviving forms. It tells of that hero's many adventures and ends with a flood story, in which a man named Utnapishtim is told by a god to build a great covered boat, and to bring into it all kinds of animals. After seven days and seven nights of flooding, Earth is covered with water. The boat comes to rest on a mountaintop.

Figure 14.4 Top: William Loftus's 1850 sketch of the Great Ziggurat of Ur, first build ca. 21st century BC. Bottom: Reconstruction, based on the 1939 drawing by Woolley.
Source: Wikipedia Commons

Utnapishtim sends birds from the boat, which return, until finally one does not, indicating the water has receded.

Sumer used sexagesimal (base 60) numbers, which we retain today in measuring degrees of a circle, minutes, hours, days, months,

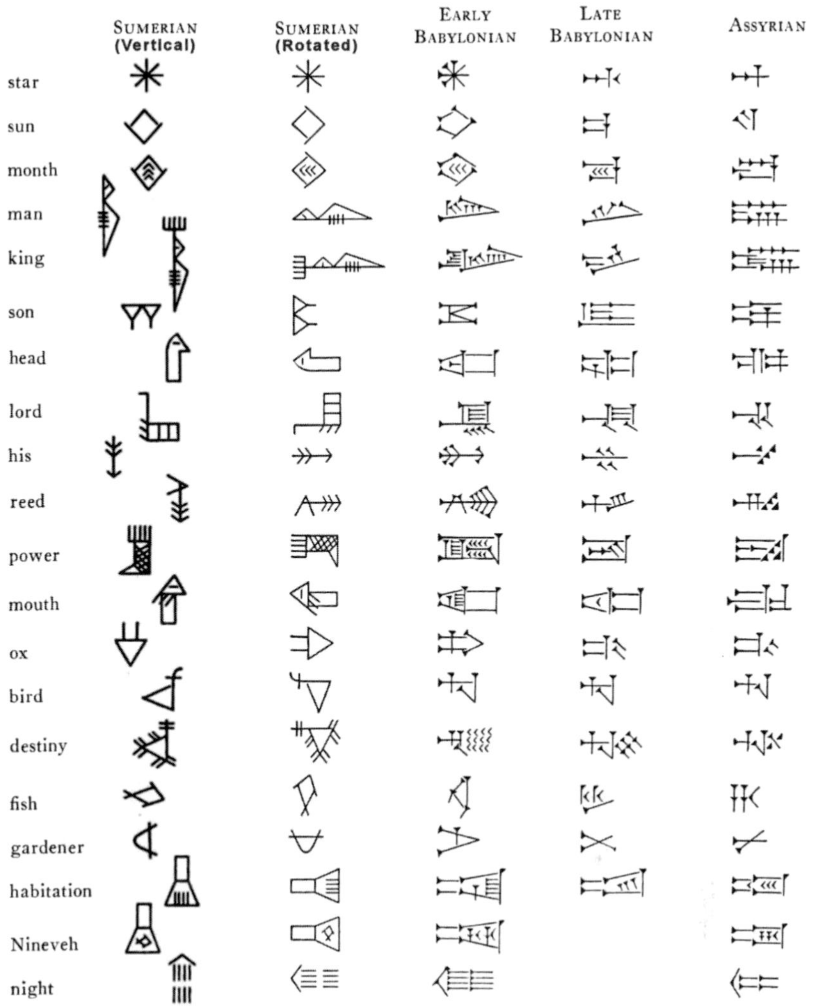

SUMERIAN (Vertical)
SUMERIAN (Rotated)
EARLY BABYLONIAN
LATE BABYLONIAN
ASSYRIAN

star
sun
month
man
king
son
head
lord
his
reed
power
mouth
ox
bird
destiny
fish
gardener
habitation
Nineveh
night

Figure 14.5 Cuneiform evolution from archaic script, based on rotated characters, by William Mason.
Source: Wikipedia Commons

and approximate years. Other innovations were the plow, canals, pottery wheels, and wheeled wagons. Imported copper was alloyed with arsenic or tin to make bronze, which could be melted and molded into any desired shape.

The city states of Sumer sometimes cooperated, sometimes competed, the dominant one shifting from time to time, usually Uruk or Ur. In 2400 BC, further to the north, a man named Sargon formed the short-lived Akkadian domain, perhaps the first great empire, to be succeeded by ever larger empires: Babylon (eighteenth century BC), which introduced King Hammurabi's Code of laws, mostly retaliatory measures specifying how retribution for one injury deserved a counter injury ("An eye for an eye"), then the Hittites (sixteenth century BC) who became militarily dominant by introducing domesticated horses and chariots, though they may not have been the inventors. Eventually the Assyrians became the leading power, reaching their height in the tenth to seventh centuries BC, bringing us into biblical history. The strong Assyrian army with archers, catapults, and siege machines destroyed the Hebrew kingdom of Israel, giving rise to the legend of the "ten lost tribes of Israel." The Assyrians in turn were overwhelmed by a revived and much greater Babylonian empire with grand architecture, still of mud brick but now beautifully colored and producing such master works as the Ishtar Gate (Figure 14.6), the fabled Hanging Gardens of Babylon (one of the Seven Wonders of the World), and a towering ziggurat seemingly the inspiration for the

Figure 14.6 Ishtar Gate of Babylon, faced with blue tiles, ordered by King Nebuchadnezzar II, about 575 BC, reconstructed at the Pergamon Museum in Berlin.
Source: Wikipedia, Creative Commons Attribution 2.0 Generic license

Bible's Tower of Babel. Babylon is also famous in biblical history for conquering the remaining Hebrew kingdom of Judah, its capital in Jerusalem, bringing captive Jews back to Babylon. After the fall of Babylon in 539 BC, the victorious Persian king Cyrus the Great allowed the captured Jews to return to Jerusalem. Persia was the new hegemon until conquered by Alexander the Great of Macedon.

We can halt this rush though Near Eastern empires, having nearly reached the beginning of the soon-to-be colossal Roman Empire. In the meantime, the Mesopotamian states had developed extensive travel networks, including ships visiting ports on the Persian Gulf and across the Mediterranean, and land routes for trade and military exchanges with Egypt.

Original Civilizations

The emergence of agrarian-urban civilization began about five millennia ago. In one of the most remarkable of coincidences, it occurred not once but in at least six places within the next 3,000 years: Mesopotamia, Egypt, India, China, Mesoamerica (i.e., central Mexico down into Central America), and Peru. Early civilization was not limited to these six, but these are the best known and probably earliest in their respective parts of the world, and they seem to have developed more or less independently of one another, becoming centers of diffusion to nearby regions.

In 1968, as this surprising picture of near simultaneity was recognized, the Swiss author Erich von Däniken argued in his book *Chariots of the Gods* that extra-terrestrial visitors must have planted the seeds of civilization around the world. To justify this fantastic claim, he asked what else could explain civilization's sudden emergence on so many continents?

There is a tendency to oversimplify matters by assuming that each transformation was a sudden and invariant sequence. In fact, the changes occurred over many generations, and there are differences among the core areas as striking as their similarities. The pristine civilizations of Asia all occur in major river valleys, which is not true of America. Egypt was much less urbanized than Mesopotamia. The appearance of any one cultural element is variable, for example, the use of animals and the wheel for transportation was important in the Old World but not in the Americas; metallurgy has appeared in some pre-civilized cultures but not in all civilized ones. Still, one can hardly fail to be impressed by the similarity that did occur or to wonder why it happened.

Figure 14.7 The Stepped Pyramid of Djoser, twenty-seventh century BC, precursor of Egypt's great smooth-sided pyramids, resembles a Mesopotamian ziggurat. *Source*: Charles J. Sharp, Creative Commons Attribution-Share Alike 3.0 Unported license

To the modern traveler, Mesopotamia and Egypt seem too close to have developed their civilizations independently. Exchange between the two areas occurred in the fourth millennium BC, as shown by Mesopotamian-styled artifacts at Egyptian sites, but it broke off within two or three centuries, the reasons unknown, leaving it unlikely that Egypt imported the basics of its later civilization, though it is tempting to wonder if pyramids, especially the earliest or "Stepped Pyramid" (Figure 14.7), was somehow inspired by the ziggurat. This does not preclude independent invention. After all, Mesoamerican cultures also used ziggurat-like temples and pyramids, including the humungous Pyramid of the Sun (Figure 14.8), which was built much later, ca. 200 AD, though still well before any contact with Old World civilizations.

Explanations

Hardly anyone today seeks one simple explanation for why certain agrarian settlements developed into civilizations. I prefer to deal with it piecemeal, looking at factors that may have predisposed

Figure 14.8 The Pyramid of the Sun, Teotihuacan (Mexico), ca. 200 AD. Mariordo, under the Creative Commons Attribution-Share Alike 4.0 International license

different groups of people to innovate in more or less the same direction (Mazur 2007).

The first factor to consider is the nature of the *sapiens* mind, as acculturated during the early Neolithic, by then having essentially modern cognitive and linguistic skills. This is a touchy matter, often connoting sexist biases. For instance, we persistently see the power of ancient civilizations residing in kingships rather than queenships, suggesting some natural tendency for males rather than females to head communities, notwithstanding an occasional matriarch. Some biological basis seems likely for this male leadership, which exists in nearly all cultural settings as well as among nonhuman primates; however, it may be nothing more than the greater size and strength of men, giving them an advantage as warriors. If so, it follows that men lose this advantage when community power is detached from physical strength, which is presumably true of modern society. Although men might still rule by tradition, there would be no intrinsic reason to regard this gender bias as inevitable.

The tendency to favor one's dependent children is part of mammalian nature, and it makes good evolutionary sense, for parents without this regard would leave few offspring in succeeding generations. At least for humans, there is an emotional mechanism at work here, acting through the affection and identity that parents feel for their children, which intensifies the more they interact with one another. If the children are happy, then usually the parents are happy; if the children suffer, then the parents suffer vicariously. That is why parents take special pains to ensure the wellbeing of their offspring, and it likely explains the reliable presence of hereditary dynasties – the passage of power and privilege to one's children – not only in ancient societies but in many societies today.

We feel special concern not only for our children but for friends and close relatives too. Humans have an emotional tendency to like and identify with anyone with whom we interact positively, whether genetically related or not. That is why families and colleagues form cliques for self-interest, especially as the population of a community grows to the point where many people are strangers to one another. It explains why my brother and I will set aside our own argument to coalesce against a neighbor who threatens one of us, and why we and the neighbor will set aside that dispute to join forces against a foreign community. If one community captures another, it explains why the winners hold themselves separate as a group from the losers, taking their wealth and forming an incipient class system. All of these features of political behavior would be expected to occur in any human society with a population exceeding the size where everyone knows everyone else personally.

Language, the substratum of human culture, carries its own imperatives. As one example, consider the findings of a classic study of basic color terms in ninety-eight languages representing a wide variety of linguistic stocks (Berlin and Kay 1969).

1. All languages contain terms for white and black.
2. If a language contains more than two color terms, it contains red.
3. If a language contains four terms, then it contains either green or yellow but not both.
4. If a language contains five terms, then it contains both green and yellow.
5. If a language contains more than five terms, then it has all of the above terms plus blue.

The order of inclusion of color terms is white and black, red, green or yellow, and blue. Other language researchers have since noted exceptions, but the findings still seem to hold as a rough generalization. What explains the striking consistency of this language pattern? Why should red be more common than brown (the color of wood, sand, and many soils) or gray (dusk, clouds)? There are vegetated environments where a term for green does not appear and desert environments where it does.

The solution apparently lies in the physical structure of our eye. The human retina contains two types of light receptor cells: rods and cones. Rods are more sensitive than cones to light, but they recognize no colors, only variations between white and black. There are three kinds of cones, each maximally sensitive to either red, blue, or green light, with red cones the most sensitive and blue the least. If we use this degree of sensitivity as an ordering principle, then the prime colors are ranked white and black, red, green, and blue. We have missed yellow, which has no special cone and remains unexplained. Otherwise, the physiology of the eye accounts for the ordering of color terms in language. We might expect a similar pattern of language terms for taste, since the taste buds of our tongue are tuned to sweet, sour, salt, and bitter.

Possibly the human mind has intrinsic preferences for certain shapes and textures, say, circles over squiggly enclosures, the large eyes and soft skin of a baby, the appeal to men of breasts and buttocks of a young woman, the appeal of shiny gold, an aversion to slimy crawling things, the avoidance smell of human (but not horse) feces. That roughly pyramidal or ziggurat-shaped structures were used for huge monuments by different civilizations, despite the impossibility of all being imitations of one original, may be due to some intrinsic appeal of this form, like a Jungian "archetype." Or it may reflect the natural physics of available building materials. A pile of sand on the beach takes a similar conical shape. It is an obvious solution to the problem of building massively tall but still stable conglomerates of earth or stone, which in our age of steel and reinforced concrete is no longer an imperative.

Environmental Biasing

People learn from their environments, and to the extent to which all humans experience their environments in the same way, different societies can learn similar things. Thus, from common experience, every human group recognizes that men and women are different classes of people, that babies come only from women, and that women have a special role in nurturing infants. Men and women desire to copulate and figure out how to do it even without tutelage, and usually understand this to be a prerequisite for pregnancy.

The sun is the most salient physical object in the environment, visible to everyone. Any human society would be concerned with the sun and its obvious characteristics, including the color yellow – the one color term not explained physiologically – as well as its circular shape, its warmth, and its cyclic coming and going, bringing light and darkness. If there is a human tendency to worship, then the sun would be a good candidate for veneration.

It is nearly inevitable that people all over would mark time by the coming of the dawn or darkness – something like a day. It is nearly as likely that people would see the changing phases of the moon as another natural time period, calculating it at about thirty days (actually 29.5), the approximate length of the menstrual cycle. People would be aware that the lunar month is shorter than the cycle of changing seasons, whether the alternation of rainy and dry spells near the equator or the movement from summer through winter and back to summer in the temperate zones.

People living in temperate zones (but not on the equator) would notice that winter days are shorter than summer days, which is why winter is cooler than summer. Today we know this occurs because Earth's orbit around the sun is tilted from the plane of the equator. The shortest day of the year (about December 21 in the Northern Hemisphere) occurs when Earth reaches the "highest" point on its orbit. This is the winter solstice, which was often marked by a special festival among ancient people and seems to be the basis for setting Christmas in December. Similarly, the summer solstice occurs on the longest day of the year (about June 21) when Earth is at the "bottom" of its orbit. These significant events were well known to ancient peoples, as they would be to anyone who routinely watched the sky. Ancient monuments, famously Stonehenge, are oriented to catch the sun at the solstices.

In every civilization, people knew that the year, the cycle of seasons, was about 365 days. (Some calendars used for ritual purposes are based on other counts, for example, the 260-day year of the Aztecs or the 354-day Muslim year, but astronomers of each civilization were aware of the 365-day cycle that correlated with the seasons. On the equator, two cycles of rainy and dry seasons make up one year.) This seems an amazing feat to modern people who do not often watch the sun rise and set on the horizon, but to a people who have that luxury, it is a simple matter to calculate the number of days in a year. In temperate latitudes the sun rises (and sets) at a slightly different point on the horizon every day, which is easily marked by such landmarks as a distant peak.

This set point moves northward as the date approaches the summer solstice, reaching its northernmost point on the solstice itself, after which the point of sunset begins moving southward again, reaching its southernmost point at the winter solstice. To calculate the length of a year, all one need do is count the days it takes for the setting sun to return to one of these extreme points, which is about 365 days (Aveni 1989).

I have gone into some detail here to illustrate how probable, if not inevitable, it is that any people who could see the sky and count into the hundreds would arrive at a calendar based on days, months, and a 365-day cycle of seasons. A civilization with specialized astronomers would be carried further along this path, eventually predicting eclipses and wondering if other events on Earth were predictable from configurations in the night sky (see Chapter 5).

The pristine states of the Old World, but not of the New, arose along great rivers whose seasonal floods provided fresh nutrients to the soil as well as a convenient water source that could be managed by irrigation. Variations in these local environments no doubt affected development in each area. The Nile gave the Egyptians some advantages over the Mesopotamians, its annual flooding more regular and dependable than that of the Tigris and Euphrates. Also, whereas the Mesopotamian plain was broad and open to invasion from all sides, the Nile Valley was protected on the east and west by desert, being exposed only from the north and south. Thus, invasions were less frequent and could be dealt with by a unified army. Furthermore, since the Nile's fertile strip was rarely more than ten miles wide, communities were strung out along the river and therefore easy to control. Probably because of this linear dispersion of the population and their little need for defensive walls, the Egyptians never aggregated into cities as large as those of Mesopotamian civilization (Wilkinson 2010).

Cities were and remain the common centers of civilizations. They seem to be inevitable when populations at that level of culture grow in numbers. Spanish conquistadores encountered magnificent New World cities like Tenochtitlan of the Aztecs and Cuzco of the Inca. These were in many ways familiar because they were like cities they knew from home. Despite there being no prior contact, each place had evolved its own cities independently.

> Shaped by cognitive impulses that are universal, people created the exact same template for crowd-based living, with

neighborhoods, open spaces, monumental architecture, and housing whose sizes and styles varied according to the social status of the occupants. Cities always have the same components: streets and neighborhoods, markets and government buildings, open spaces and crowded alleyways. There are places for entertainment and places for education. The necessary connectivities of infrastructure, like water supplies and roads and bridges, are exhibits of sober engineering but also a celebration of soaring architectural achievement . . . There are areas in which the richest people reside, and there are slums in which the poorest take shelter. There are merchants and bookkeepers and teachers who constitute a middle stratum of rank and wealth whose houses and possessions reflect a preoccupation with status and comfort as well as a capacity to buy the little extras of urban décor. And although the religious tradition is different from place to place, people clearly demarcated the sites of ritual activity with monumental building and grand surrounding spaces. (Smith 2019: 24).

Indus River and China

Most mysterious of the original Bronze Age civilizations was one along the Indus River. It is sometimes called "Harappan civilization" after its largest city, Harappa. Charles Masson (1800–1853), a deserter from the East India Company's army, was the first European to see the city's buried ruins, describing them in 1842, but he did not recognize their antiquity or importance. In the early twentieth century, still unaware of the site's significance, European archeologists excavated Harappa and another city, Mohenjo-Daro (Figure 14.9), finally realizing they had found an ancient long-forgotten civilization. Continued work uncovered three more major urban centers and over a thousand smaller cities and settlements (Singh 2009; Wright 2009).

Indus civilization's mature period, dated 2600 to 1900 BC, was contemporary with Mesopotamia and Egypt. There are indications that Indus boats traded with Mesopotamia via the Persian Gulf, but the civilizations are considered too dissimilar to have a common origin.

Indus civilization occupied more area than Egypt or Mesopotamia, covering most of today's Pakistan and northwestern India. Unlike the

Figure 14.9 Excavated ruins of Mohenjo-Daro, on the right bank of the Indus River, Pakistan, showing the Great Bath in the foreground. UNESCO World Heritage Site.
Source: Saqib Qayyum, under the Creative Commons Attribution-Share Alike 3.0 Unported license

others, it lacks monumental works, no palaces or temples, suggesting a more egalitarian population. The cities are especially impressive for their advanced urban planning, with well-built mud-brick houses and other structures lining straight streets, provided elaborate water supply, drainage, and waste disposal systems.

There were high levels of technology and art, including new methods of crafting in ceramics and metalworking; great accuracy in measuring weights, measures and time using the decimal system; and lifelike sculpture and statuettes. Beasts of burden pulled wheeled wagons, as shown in miniature reproduction, apparently a toy (Figure 14.10).

Indus cities became depopulated around 1300 BC and eventually forgotten. There is no indication of conquest or destruction. A common speculation is that drying climate made the food supply unsustainable, forcing inhabitants to migrate. Even now, we know little of these people, their language, where they came from, or where they

Figure 14.10 Toy cart pulled by animals from cache at site of Daimabad, 2000 BC.
Source: Author Miya.m, Creative Commons Attribution-Share Alike 3.0 Unported license

went. Inscriptions that seem to be writing have not been deciphered, and there are few skeletal remains, none so far allowing DNA analysis. Peter Bellwood suggests that the Harappan moved into the Ganges basin and became the early historical civilization of India, with Indic languages and early Vedic Hinduism, though this is unconfirmed.

The Himalayas and the Tibetan Plateau separate India from China, where one more pristine state arose, now fairly claiming to be the oldest continuous civilization of all. Early Chinese documents describe a Xia Dynasty (ca. 2070–1600 BC) but there is no independent evidence of it. The first verifiable dynasty is the Shang (ca. 1600–1046 BC) in the Yellow River valley, though perhaps there were Yellow and Yangtze river civilizations millennia earlier, the uncertainty reflecting a dearth of archeological explorations. Since the Shang, Chinese history is commonly enumerated by successive dynasties, which despite some wrenching disruptions continued until the last emperor, a little boy named Pu Yi, was overthrown by revolution in 1912, ending his days as a gardener under Maoist Communism.

The Zhou Dynasty (1046–256 BC), following the Shang, introduced the idea of the Mandate of Heaven, that the emperor was naturally the just ruler of the Celestial Kingdom. If he were overthrown, that was understood to mean he had lost the mandate, which was transferred to his successor – something occurring fairly often. Conquests, famines, and natural disasters might be taken as signs that the present emperor had lost the mandate as a just ruler (Murowchick 1994).

In 221 BC, Qin (pronounced *chin*) Shi Huang conquered and unified warring splinter groups, founding a dynasty that lasted only fifteen years, leaving behind the eponymous name China and a spectacular clay army buried with the emperor for his use in the afterlife. This Terra Cotta Army, discovered in 1974 by farmers digging a water well, turned out to be part of a much larger necropolis that required an enormous number of laborers including skilled craftsmen. The site, near the city of Xi'an, not yet fully excavated, contains thousands of life-size, realistic, individualized warriors, once painted but now faded, as well as horses and chariots. The array is stunning (Figure 14.11), rivaling the Great Wall as a modern tourist attraction.

The Han dynasty (202 BC–20 AD) followed the Qin and is regarded a golden age for its extended period of stability and prosperity, consolidating China as a unified state under a central imperial bureaucracy,

Figure 14.11 Pit one, 230 meters (750 ft) long and 62 meters (203 ft) wide, contains the main army of more than 6,000 figures. Wooden ceilings were covered with reed mats and layers of clay for waterproofing, and then mounded with more soil.
Source: Wikipedia Creative Commons. A UNESCO World Heritage site.
Source: Jmhullot, under the Creative Commons Attribution 3.0 Unported license

and its improvements in arts, science, and technology, including papermaking. Confucianism became orthodox philosophy, stressing the importance of the family, order, and social harmony, obedience to elders and people of authority, especially the emperor. Subsequently most Chinese people call themselves Han, an exclusive ethnic label.

The Silk Road, so named from the lucrative trade in silk carried out along it length, was a network of routes begun by the Han to cross what had been an isolating barrier into central Eurasia. The Great Wall of China (Figure 14.12), actually several walls often extended or reconstructed, was built and maintained by many dynasties including the Han. The stone version familiar to us today was a later construction of the Ming Dynasty (1368–1644), eventually extending 4,500 miles. (It is not true that astronauts see it from space.) Its main purpose was to protect against invaders from the north, though it never did successfully work for that because invaders were able to bribe their way though. Another purpose, more successful, was to control entry into China for collecting customs taxes.

Figure 14.12 The author's wife with an unidentified companion on the Great Wall, near Beijing on an unusually clear day

Comparing the Roman and the Han Empires

For four centuries the Roman and Han empires were the two greatest powers in the world, yet were only dimly if at all aware of each other (Aldrete 2011). Rome was founded earlier but did not reach full dominance of the Mediterranean until decisively defeating Carthage in 202 BC, the same year the first Han emperor came to power.

Both empires had nearly equal populations, together including perhaps half the people then in the world. Their land areas were similar though shaped differently. Roman territory was coastal, encircling the Mediterranean Sea, while Han territory was compact within its physical boundaries (Figure 14.13).

Facing the challenge of administering territories and populations of unprecedented size, both empires used similar means to do it, and were remarkably effective within the limits of the time. They built excellent networks of roads, bridges, and canals to expedite transportation and trade and the movement of armies, and they created courier or postal services to carry messages over these long distances. Roman roads were built by the army. Its first task on conquering a new territory was linking it to the road network. Chinese roads were built by conscripted peasantry. Apart from the army role in roadbuilding, Rome depended heavily on slave labor, much more so than China.

Even with good roads, land speeds rarely exceeded twenty miles per day, reaching 100 miles with extreme effort. Travel by water,

Figure 14.13 The Roman and Han empires on opposite sides of Eurasia, ca. AD 1. (Partially based on *Atlas of World History* (2007) – World 250 BC–1 AD).
Source: Gabagool, under the Creative Commons Attribution 3.0 Unported license

depending on variations in weather, was slower but might take more direct routes and had the great advantage of cheaply carrying bulk cargo including foodstuffs. Bandits and pirates plagued travel and trade in both empires; their control was a central goal of administration.

Roman engineering was of particular excellence, some roads and bridges still usable today. Giant Roman aqueducts remain standing. Much of this expertise disappeared with the decline of the empire. One example is the ability to use cement that hardened under water, an enormous asset in building harbors. The technique was lost during the Middle Ages, not recovered until recent centuries.

Han engineers were no slouches. Their roads were good, if not of Roman quality, but their canals were excellent. China's two great rivers, the Yellow and Yangtze, run east and west. Lacking north-south water transport for bulk cargo, the earliest dynasties dug canals. Large sections were built under the Han, but they were not linked until the Sui dynasty (581–618 AD), creating the Grand Canal, still the longest artificial river in the world, running 1,104 miles (1,776 km). Periodically restored and rebuilt, it is still in use as a major waterway (Needham 1986).

Control of outlying territories was difficult for both capitals. An order or inquiry might take a month to reach its recipient, assuming it successfully completed the journey, and double the time if a response was required. Of necessity, each empire was divided into many administrative provinces, those farthest from the capital under less control, where decision-making was mostly left to local governors.

Both empires ruled through formal bureaucracies, but they differed in size and form.

The Han had a rigid formal hierarchy, 120,000 career bureaucrats arranged in twenty grades or ranks, each with specific duties, privileges, and signs of status. Entry was through civil service examination based on Confucian classics, requiring extensive preparation. In principle open to anyone, in fact only the affluent could afford the instruction needed to qualify. Once qualified, an official would strive to move up the ranks.

The Roman bureaucracy was more ad hoc and had far fewer official positions, less than 200 senior administrators. These were aided by informal agents including friends, slaves, and freemen who carried out most of the daily chores. One outcome similar in both systems was that formal administration was in the hands of landowners. To reduce corruption, officials were assigned to areas far from their own homes.

As Romans expanded their territory by conquering new lands, it was their practice to govern through indigenous authorities. Herod of the New Testament is an example, well known for his role leading to the execution of Jesus Christ. Herod was a Jew, ruling as the local surrogate for Emperor Augustus. The Romans took pains to assimilate leading outsiders into their own culture, as was Herod, who became a great builder of Roman-style constructions, still seen across modern Israel. The Han were more likely to impose their own bureaucracy on a newly acquired territory, but like the Romans, sought to inculcate local elites into the dynastic culture.

Both empires had prevailing ideologies, which may or may not have contained religious elements. Following the Qin, legalistic Confucianism was the ideology of the Han, continuing throughout dynastic history, even after joined later by Taoism and Buddhism as foundation beliefs. Furthermore, Chinese society, apart from hierarchy, was and continues to be relatively homogeneous. Though languages spoken across China were not mutually intelligible, in written form they were.

The Romans were not so particular about conformity. Latin was the official language, used by the elite, but many other languages were used and considered acceptable. They tolerated many cults and religions, if members caused no trouble. Even when Emperor Constantine (272–337 AD) adopted Christianity, this was not obligatory for others though gradually it supplanted pagan religious forms across the empire.

The Roman elite idealized and emulated Greek culture, despite having conquered Greece. This is evident in their initially adopting the Greek pantheon of gods and heroes, and in the continuity between Hellenic and Roman art and architecture.

Both empires viewed themselves as superior to any other, bringing their subject peoples economic prosperity and stability, and civilization to barbarians, though also taxation and sometimes conscription. Both were perennially threatened by barbarian tribes, requiring large armies and extensive defensive walls. Both suffered from serious bouts of disease, some illnesses derived from their livestock. Both sometimes suffered from some weak, inept, and even insane leaders, though it is not clear that these changing qualities had important impact far from the capital.

During the early part of its period of greatest power, Rome was a republic meaning that some men were citizens, a status that conferred certain rights and obligations. At least within that limited group, there was popular government, institutionalized in the Senate. Thus, for some Romans there was a sense of individual entitlement, which persisted even after 27 BC when the government became imperial. This gave the Roman populace a voice in public affairs never held in China. Roman emperors, even with absolute power, had to maintain the good will of the people, to be seen by them, appearing on coinage, portrait busts, and in person at public entertainments such as were held in the Coliseum. Appearance was important. Not so in China, where emperors were not seen, sequestered from public view, at least if the dynasty remained in power.

Both empires eventually failed due to external threat, internal dissention, economic collapse, new religious movements, and endemic disease. Both split in two, one part taken over by barbarians. In China, the change in dynasties was quick. For Rome, the decline took centuries.

Like the similarities the conquistadores found between cities in Europe and America, which had no prior contact, the histories of the Roman and Han empires suggest a certain inevitability to the course of development. On the other hand, subsequent developments on opposite sides of Eurasia show the different directions that can be taken.

Europe and China Subsequently

The legacies of Rome and the Han survived their fall, obviously in the Eastern Roman (Byzantine) Empire, which persisted until its defeat by Muslim forces. Its capital on the Bosporus, Christian Constantinople, was conquered by the Ottomans in 1453, becoming Islamic Istanbul, with churches converted to mosques.

Islam is an "Abrahamic" religion like Judaism and Christianity, so called because all believe in the biblical patriarch Abraham, regarding him as their common progenitor. All believe in the same one god (but not the Christian trinity), whom Jews call Yahweh and Muslims call Allah. Despite their kinship, the faiths have a persistent proclivity toward incivility, not least the several Crusades for control of Jerusalem, a holy site for all three religions.

The beginning of Islam is dated to 622 AD, when Mohammed and his few followers fled from Mecca to Medina, both cities on the

Arabian Peninsula. In following years, the new religion grew explosively and militantly as Arab armies conquered lands and converting people through much of the Middle East, North Africa, and the Iberian Peninsula. Islam would begat its own sparkling civilizations, nourishing learning, arts, and architecture, of which one early example is Jerusalem's magnificent Dome of the Rock (Figure 14.14). It encases the rock from which Mohammed rode his horse up to heaven, coincidentally, the same rock upon which Abraham, on God's command, nearly sacrificed his son Isaac, until reprieved at the last moment.

Antipathy between Christendom and Islam eventually led medieval mapmakers to convert Europa, once a region on Roman maps, into *Europe*, turning a peninsula of Eurasia into a new continent.

For Europeans, the centuries of perceived decline from the cultural height of the Greco-Roman world until the Renaissance revival of classical works was naturally enough called the "Middle Ages," a term now more often referring to the centuries in Europe separating the Classical Age from the Modern Age. The view of these middle centuries was one of stasis if not regression.

> The impression is created by emphasizing the slow pace of technological change, the closed character of feudal society, and the fixed theocratic perceptions of human life. The prime symbols of the period are the armoured knight on his lumbering steed; the serfs tied to the land of their lord's demesne; and the cloistered monks and nuns at prayer. They are made to represent physical immobility, social immobility, intellectual immobility (Davies 1996: 291).

Modern historians disdain the once accompanying term "Dark Ages" as too prejudiced against non-Christian cultures of barbarians and Muslims, also perhaps understating accomplishments that did occur in Europe during this time. A distinction is now made between the early and later Middle Ages, before and after the eleventh century. The earlier period is more clearly one of decay compared to the relative effervescence of the later one. Whatever one calls this pause, it was limited to Christian Europe and not reflected elsewhere.

From the perspective of the history of science and technology, there was indeed in the early Middle Ages a glaring comedown from Roman engineering and Greek science, with European thought and

Figure 14.14 The author's granddaughter, barely seen next to her large father, in front of the Dome of the Rock, completed 692 AD, atop the Temple Mount in the Old City of Jerusalem. The dome is gold colored and the building faced with blue tile.

literature, even the ability to read and write, turning to the Church. Thankfully, the new Arab civilizations became repositories for these scholarly works by the Greeks and Romans, translating them into Arabic and making improvements. By the twelfth century, they were being sought by Europeans and brought home for translation into Latin, seeding an intellectual reawakening.

Whatever one calls this European pause, it may have allowed China a lead in technology. Beside the famous innovations of paper-making, the compass, gunpowder, and the printing press with moveable type, China was advanced in many areas of invention from banknotes, blast furnaces and bombs to well drilling, wheelbarrows, and wrought iron, explicated in a remarkable series of books by historian Joseph Needham on ancient Chinese science and technology. In the fifteenth century, when Columbus "discovered" America, China had as good a claim as any of being the most advanced civilization in the world. The Forbidden City of the Chinese emperors as well as the Islamic world of the Moors were as grand as anything in Christendom.

Voyages of Discovery

The sea routes to Asia and the Americas that Europeans discovered in the fifteenth and sixteenth centuries brought huge wealth that was invested in new trade and a general growth of commercial activity. More importantly, they led to a vast overseas system of colonies, which eventually supplied raw material for European factories and the markets for finished goods. Ironically, the leader in overseas exploration was Portugal, one of the last nations of Europe to industrialize.

Portugal was ideally situated for voyaging into the Atlantic. Hemmed in by Spain and without a coast on the Mediterranean, there was no other option. Prince Henry the Navigator (1394–460) made exploration a national mission, his purposes to increase Portuguese commerce and combat the spread of Islam (Boorstin 1983). Year after year, Portuguese ships explored the west coast of Africa while developing a lucrative trade in slaves, gold, and ivory. Finally, in 1497–1499, Vasco de Gama (ca. 1460s–1524) sailed around the Cape of Good Hope into the Indian Ocean, returning with a valuable cargo of spices from India. Da Gama went again to India with a military squadron, killing and looting to establish by force a Portuguese commercial empire throughout the Indian Ocean, with trading outposts eventually reaching to China and Japan.

In the meantime, Christopher Columbus (d. 1506) was seeking a monarch to support his search for a western route to the Orient. Portugal was preoccupied with its eastern route, but Isabella of Spain, fresh from her successful expulsion of the Moors and Jews, financed the project as a means of spreading Christendom. He made four voyages to America (1492–1504), always insisting that he had reached the Orient.

More ships quickly followed the great discoverers. The Portuguese consolidated their eastward route around Africa but also sailed west to colonize Brazil. The Spanish searched for and found treasure in the New World. As it became clear that America was not Asia, Columbus's plan was revitalized, for how much farther could China be? Spain sent Ferdinand Magellan to see. In 1519, Magellan (1480–1521) set off on the greatest voyage ever, leading five ships across the Atlantic, then down the coast of South America, threading them through the excruciating strait at the bottom of the continent that now bears his name, emerging into the Pacific Ocean. He then crossed that unknown sea, a feat taking three and one-half months (compared to five weeks for Columbus to cross the Atlantic), and finally reached islands off Asia. While in the Philippines, Magellan became needlessly involved in an intertribal fight and was killed by natives. The one ship that was still seaworthy continued westward, rounding Africa and returning to Spain three years after it had departed, with 18 men left from the original 250, thus completing the first circumnavigation of the globe.

Columbus was not the earliest European to reach America. Vikings had mapped Canada near Hudson Bay in the eleventh century, and there was a Basque whaling port in Labrador in the 1500s, but these were temporary presences (Tuck and Grenier 1985). The Spanish mission to spread Christianity, also to plunder and acquire slaves and plantation land, was very lucrative and highly destructive, subjugating the American civilizations and spreading European diseases that wiped out most of the indigenous people (Mann 2006, 2012).

As Spain exploited America, Portugal forcefully colonized the Indian Ocean. These two nations greatly increased the commercial activity of Europe and its concentrations of wealth, shifting the center of business activity from Italy and the Mediterranean to the Atlantic Coast. Other coastal nations, seeing the benefits of overseas possessions, were spurred into their own voyages of discovery and colonization, especially France and England, and briefly the Netherlands. The perennial warfare and shifting alliances of Europe were now extended to the distant colonies and oceans, with the result that Portugal soon lost its commercial empire in Asia to the Dutch.

The details of European politics need not concern us here, but it is worth noting the increasing stature of England, a beneficiary of the New Atlantic commerce. That island nation had naturally developed as a sea power, but it lacked colonies. Its first New World profits came from raiding Spanish treasure ships returning from America. This and other factors led Spain to send a mighty naval armada against England in 1588, but it was utterly destroyed by a combination of bad luck, bad planning, and superior English seamanship.

Shortly afterward, England (like France and the Netherlands) began seeking its own footholds in India and America. The English colonies in North America did not produce gold and silver but did provide plenty of land for agricultural products that could be sent home. The colonists, often religious pilgrims, had come to stay, forming permanent settlements and ultimately an independent nation (Mazur 2007).

A Counterfactual Fantasy of Chinese Dominance

Eighty years before Columbus's voyages, during the Ming Dynasty (1368–1644), the eunuch Admiral Zheng He (1371–1434) commanded Chinese treasure fleets of sixty or more ships on seven voyages through the South China Sea into the Indian Ocean, even visiting the east coast of Africa. The purposes of these voyages were to trade, promote diplomatic relations, and to impress the places visited with China's imposing power and to fortify its collection of tributes. Admiral Zheng was also effective in stopping pirate raids along these sea lanes.

His fleets were of unprecedented size, if we are to believe available sources, but their routes were not completely new as Chinese ships had long visited Arabian ports. Zeng's own ship was enormous, reportedly having nine masts and being four times the length of Columbus's flagship, *Santa Maria* (Figure 14.15). Under a new Ming emperor, Zeng's voyages were halted, as was ship building and ocean voyaging generally. The voyages seem to have been successful, though no doubt costly. The reason they were stopped is not clear, but it is likely they took second place to domestic priorities.

What if the Chinese had "discovered" the New World before Columbus? It would have necessitated a longer voyage, but Magellan accomplished that. What if the Chinese, not the Spanish, had been first to meet and conquer the Aztec, Mayan, and Incan armies?

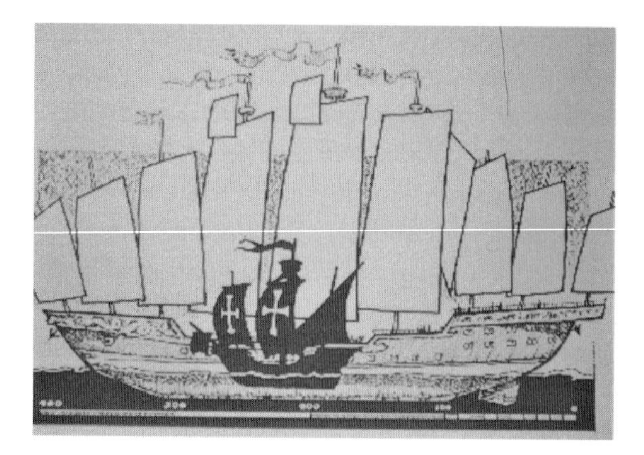

Figure 14.15 Comparison of Zeng He's flagship and Columbus's *Santa Maria*. Copyright unknown

The consistent success of small corps of Spaniards to defeat huge Native American armies is certainly impressive. Jared Diamond (1997) attributes this to a combination of "guns, germs, and steel," the Spanish muskets, swords and lances being far superior to non-metallic American weapons. Horses brought from Spain gave them an imposing advantage over native foot soldiers, and of course, the greatest killer was the European diseases that quickly spread though indigenous populations with no immunities to them. (The Europeans may have acquired syphilis from America, but this is uncertain.) Another major help was the Spanish alliances with the empires' subject peoples who hated their rulers, countering the numerical imbalances.

The Chinese, had they met these civilizations, would have had the same advantages, including plenty of germs derived from their close association with livestock. Chinese gunpowder weapons of the time were less effective than Spanish muskets, but their importance may be overstated because muskets were not very good either. Bernal Diaz (ca. 1496–1584), who fought with Cortes and later wrote an account of his adventures, including many battles in which he participated, consistently describes the Spanish weapons for close combat as swords, crossbows, and muskets. Indeed, Cortes's expedition set out with more crossbowmen than musketeers (Diaz 1963). The Chinese invented the crossbow some 2,000 years earlier, even developing repeating crossbows that fired several bolts. And they were famous for their horse-mounted warriors, so there is no reason to think they would not have been as militarily successful as the conquistadores. Might Latin Americans today be speaking Chinese?

It was not inevitable at the time that white Europeans, and later their American cousins, would come to dominate the world. That view would become common wisdom by the nineteenth century and is still held by some, though the modern re-emergence of China as a world superpower in the twenty-first century may extinguish it for good.

Industrial Transformation

A Martian visiting Earth in the year 1500 would have found one civilization like another. In all of them, most of the population raised crops and livestock. Cities and towns, the centers for commerce and manufacture, were interconnected by trade routes, some quite long, where land travel was by foot or animal, and sea travel by sail or oar. Communication moved no faster than a person could. Manufacturing was carried out by individuals or small groups of people using handcraft methods, with flowing water, wind, or fire as their source of inanimate power. Large monuments and ceremonial buildings were constructed of stone, wood, and dried brick, while metal was reserved for smaller objects. Land was the most valued resource, and even the richest people rarely had large (by modern standards) amounts of spendable capital. A hereditary monarch usually stood at the top of a rigid status hierarchy, supported by a favored aristocratic class (often warriors), and at the bottom was the poorer mass of peasants. Often, one's position in this hierarchy was fixed at birth and tied to a specific piece of land, as in feudal Europe and Japan, but sometimes careerists were assigned to distant posts, as in the administrative bureaucracies of Imperial Rome or China. In either case, the rights and obligations of one rank toward another were well specified (always to the advantage of the higher ranked), and an unquestioned religion and its clergy justified the existing arrangement as right and proper. Armies depended upon blades and animal power. Each civilization had impressive (by modern standards) artistic and technological achievements, and no one surpassed all the others.

If the Martian had returned in 1900, it would have seen European civilization as preeminent, followed by its direct descendant America. No longer agrarian, these societies – and to a lesser extent Japan – had become industrial civilizations, their populations moving from farms to cities, taking jobs in manufacture and commerce. Efficient factories employing large workforces, using steam power and even electricity, produced goods in far greater quantity, and often cheaper, than ever before. Finished products were sold around the world, transported by trains and steamships that returned with raw materials to feed the factories. Steel was a common building material, and electrical communication was instantaneous. Money and industrial resources had replaced land as the most valued form of capital, with many individuals and corporations having accumulated immense amounts. Some of Europe's powerful monarchies had fallen (others would soon follow) along with special privilege for their aristocracies; the United States was a democracy where every white man had the same rights and obligations. Catholicism was in decline in Europe, challenged not only by Protestantism but by a secular scientific viewpoint. The technological achievements of Europe and America were unequaled elsewhere, and no agrarian army could withstand their mechanized military forces.

In less than 400 years, the industrialized societies had become separated from the agrarian ones – and not only separate but in control. This division remains the most important one in the world today.

We have reached a point where Milankovitch orbital cycles and other natural causes of the coming and going of ice ages no longer greatly affect the development of society. Indeed, the natural return of an Ice Age, once expected in the near future, is forestalled because the artificial warming of the planet through industrial activity more than counters any natural cooling trend.

Little Ice Age

I cannot leave this topic without mentioning the so-called Little Ice Age that episodically cooled the Northern Hemisphere, and mountainous areas even in Patagonia, especially in the seventeenth and early nineteenth centuries. The causes are not fully known but appear to be a combination of regional effects rather than a globally synchronous cooling, and therefore not a true ice age. In any case, the cooling was modest, perhaps 1°C, relative to late twentieth-century levels, and certainly slight compared to the Younger Dryas or Pleistocene glaciations (IPCC 2001).

Still, the unusual cold added to the suffering caused by famines or other social disruptions, always affecting the poor more than the affluent, and it may have led to a rise in scapegoating as crowds blamed witches or Jews for bad seasons and poor crops. It inspired a genre of art and literature, giving us the fictitious Hans Brinker, skating the frozen canals of Holland (Dodge 1865).

"The year without a summer," 1816, occurred during the Little Ice Age, but it is difficult to know if that cooling trend was a serious contributor. There had been a terrific volcanic eruption of Mount Tambora in the Dutch East Indies (today's Indonesia) the year before, spewing emissions into the upper atmosphere, which circled the world, blocking sunlight everywhere.

I could close this chapter on that year's destructive events, crop failures and famines, but prefer to emphasize its felicitous contributions (Fagan 2001; Parker 2013). The haze in the sky caused magnificent red sunsets, and sometimes darker, moodier lighting. These were especially inspiring to the Romantic artists, notably J. M. W. Turner and Caspar David Friedrich.

In June 1816, the wet, uncongenial summer forced Mary Shelley, her husband Percy Bysshe Shelley, and their close friend, Lord Byron, to stay indoors at their villa on Lake Geneva for much of their Swiss holiday. To pass the time, Byron proposed a contest to see who could author the scariest story. Despite competing against two of the greatest poets of the day, Mary won, producing *Frankenstein, or the Modern Prometheus*, no doubt the most memorable artistic creation of that dismal year. Her creature has since become emblematic of the dark side of the industrial forces that were and are now shaping Earth and its climate.

REFERENCES

Adams, M. 1808. "Some Account of a Journey to the Frozen-Sea, and of the Discovery of the Remains of a Mammoth." *Philadelphia Medical and Physical Journal* 3: 120–137.

Agassiz, E. 1885. *Louis Agassiz: His Life and Correspondence.* 2 vols. Boston: Houghton Mifflin, 1886.

Agassiz, L. 1840. *Études sur Les Glaciers.* Neuchâtel: Jent et Gassmann, translated by A. Carozzi (1967).

Agassiz, L. 1847. *Nouvelles études et experiences sur les glaciers actuels, leur structure, leur progression, et leur action physique sur le sol.* Paris: V. Masson.

Aldrete, G. 2011. *History of the Ancient World.* Chantilly, VA: Great Courses.

Allen, S. 1999. *Finding the Walls of Troy.* Berkeley: University of California Press.

Alley, R. 2014. *The Two-mile Time Machine.* Princeton, NJ: Princeton University Press.

Anderson, A. 1989. *Prodigious Birds: Moas and Moa-Hunting in Prehistoric New Zealand.* Cambridge: Cambridge University Press.

Anderson, M. 2005. *Tending the Wild.* Berkeley: University of California Press.

Andersson, J. 1934. *Children of the Yellow Earth.* London: Keagan Paul, Trench, Turbner & Co.

Andrews, W., and T. Kaplan. 2015. Where the candidates stand on 2016's biggest issues. *The New York Times* (Dec. 15).

Argue, D., and C. Groves. The affinities of Homo floresiensis based on phylogenetic analysis of cranial, dental, and postcranial characters. *Journal of Human Evolution* 107: 107–133. doi:10.10/j.jhevol.2017.02.006

Arnold, C. 2018. *Trapped in Tab: Fossils from the Ice Age.* Caroline Arnold.

Arrhenius, S. 1908. *Worlds in the Making: The Evolution of the Universe.* New York: Harper & Row.

Aubert, M., et al. 2014. Pleistocene cave art from Sulawesi, Indonesia. *Nature* 514: 223–227.

Auel, J. 1980. *The Clan of the Cave Bear*. Crown.

Auroux, S. 2000. *History of the Language Sciences*. New York: Walter de Gruyter.

Aveni, A. 1989. *Empires of Time*. New York: Basic Books.

Ayala, F. 1995. Policy forum. *Science* 267: 826–827.

Balter, M. 2010. The tangled roots of agriculture. *Science* 327: 404–406.

Bar-Yosef, O. 1998. The Natufian culture in the Levant, threshold t the origins of agriculture. *Evolutionary Anthropology* 6: 159–177.

Baust, J., and R. Lee. 1983. "Population differences in antifreeze/cryoprotectant accumulation patterns in an Antarctic insect." *Oikos* 20L: 120–124.

Bell, R., and H. Seroussi 2020. History, mass loss, structure and dynamic behavior of the Antarctic Ice Sheet. *Science* 367: 1321–1325.

Bellwood, P. 2005. *First Farmers: The Origins of Agricultural Societies*. Carlton, Australia: Blackwell.

Bergstrom, A. et al. 2020a. Insights into human genetic variation and population history from 929 diverse genomes. *Science* 367: 1339.

Bergstrom, P. et al. 2020b. Origins and genetic legacy of prehistoric dogs. *Science* 370: 557–564. doi: 10.1126/science.aba9572

Berlin, B., and P. Kay. 1969. *Basic Color Terms*. Berkeley: University of California Press.

Bezdek, R., C. Idso, D. Legates et al. 2019. *Climate Change Reconsidered II*. Arlington Heights, IL: The Heartland Institute.

Blakey, R., and W. Ranney. 2018. *Ancient Landscapes of Western North America*. Cham, Switzerland: Springer International.

Bloom, H. 2004. *The Book of J*. New York: Grove Press.

Blomberg, S., and T. Garland Jr. 2002. Tempo and mode in evolution: Phylogenetic inertia, adaptation and comparative methods. *Journal of Evolutionary Biology* 15: 899–910.

Bobrovskiy, I., J. Hope, A. Ivantsov et al. 2018. Ancient steroids establish the Ediacaran fossil *Dickinsonia* as one of the earliest animals. *Science* 361: 1246–1249.

Boorstin, D. 1983. *The Discoverers*. New York: Random House.

Brainard, J. 2020. Top dino fetches top dollar at auction. *Science* 370 (Oct.): 268.

Braje, T., T. Dillehay, J. Erlandson, et al. 2017. Finding the first Americans. *Science* 358: 592–594. doi: 10.1126/science.aao5473

Brinker, L. 2015. Evolution and the GOP's 2016 candidates: A complete guide. Salon.com. Feb. 12.

Breton, P. 1992. *Niagara: A History of the Falls*. New York: Kodansha.

Broad, W. 2018. "How the Ice Age Shaped New York." *The New York Times* (June 5).

Broecker, W. 2010. *The Great Ocean Conveyer*. Princeton, NJ: Princeton University Press.

Brumm, A., et al. 2021. "Oldest cave art found in Sulawesi." *Science Advances* 13 (Jan. 2021): Vol. 7, no. 3, eabd4648. doi: 10.1126/sciadv.abd4648

Brusatte, S. 2018. *The Rise and Fall of the Dinosaurs*. New York: William Morrow.

Buchwald, J., and D. Greco Josefowicz. 2020. *The Riddle of the Rosetta*. Princeton NJ: Princeton University Press.

Burney, D., and T. Flannery. 2005. Fifty millennia of catastrophic extinctions after human contact. *Trends in Ecology and Evolution* 20: 395–401.

Callaway, E. 2013. Genetic Adam and Eve did not live too far apart in time. *Nature*. doi: 10.1038/nature.2013.13478.

Carozzi, A. 1967. *Studies on Glaciers Preceded by the Discourse of Neuchâtel, by Louis Agassiz*. New York: Hafner Publishing Company.

Chamberlin, T. C. 1897. A group of hypotheses bearing on climatic changes. *Journal of Geology* 5: 653–683.

Chambers, R. 1844, 1994. *Vestiges of the Natural History of Creation*. Chicago: University of Chicago Press.

Christie, A. 1936. *Murder in Mesopotamia*. London: Collins Crime Club.

Clarkson, C., Z. Jacobs, et al. 2017. Human occupation of northern Australia by 65,000 years ago. *Nature* 547: 306–310. doi: 10.1038/nature22968.

Clottes, J. 2011. *What Is Paleolithic Art?* Chicago: University of Chicago Press.

Coffey, P. 2008. *Cathedrals of Science: The Personalities and Rivalries that Made Modern Chemistry*. London: Oxford University Press.

Conkling, P., R. Alley, W. Broecker et al. 2011. *The Fate of Greenland*. Cambridge, MA: MIT Press.

Corballis, M. C., and S. E. G. Lea. 1999. *The Descent of Mind: Psychological Perspectives on Hominid Evolution*. Oxford: Oxford University Press.

Council on Environmental Quality. 1980. *The Global 2000 Report to the President*. Washington, DC: U.S. Government Printing Office.

Croll, J. 1875. *Climate and Time in Their Geological Relations: A Theory of Secular Changes of the Earth's Climate*. New York:

Cuvier, G. 1813. *Essay on the Theory of the Earth*.

Dartnell, L. 2019. *Origins: How Earth's History Shaped Human History*. New York: Basic Books.

Davies, N. 1996. *Europe: A History*. New York: Oxford University Press.

Davis, L. et al. 2019. Late Upper Paleolithic occupation at Cooper's Ferry, Idaho, USA, ca. 16,000 years ago. *Science* 365: 891.

Denham, T., S. Haberle, C. Lentfer et al. 2003. Origins of agriculture at Kuk Swamp in the highlands of New Guinea. *Science* 301: 189–193. doi: 10.1126/science.1085255

Desmond, A., and J. Moore. 1991. *The Life of a Tormented Evolutionist: Darwin*. New York: Warner Books.
De Queiroz, A. 2014. *The Monkey's Voyage*. New York: Basic Books.
Diamond, J. 1997. *Guns, Germs, and Steel*. New York: W.W. Norton.
Diaz, B. 1963. *The Conquest of New Spain*, translated by J. Cohen. London: Penguin Books.
Dillehay, T., J. Rossen, T., Andres, and D. Williams. 2007. Preceramic adoption of peanut, squash, and cotton in Northern Peru. *Science* 316: 1890–1893. doi: 10.1126/science.1141395.
Dingus, L., and M. Norell. 2010. *Barnum Brown: The Man Who Discovered Tyrannosaurus rex*. Berkeley, CA: University of California Press.
Dolnick, E. 2021. *The Writing of the Gods*. New York: Scribner.
Easby, D. 1965. Pre-Hispanic metallurgy and metalworking in the New World. *Proceedings of the American Philosophical Society* 109: 89–95.
Eisley, L. 1961. *Darwin's Century*. New York: Anchor.
Eldredge, N. 1991. *Fossils*. New York: Harry N. Abrams.
Eldredge, N., and S. Gould. 1972. "Punctuated equilibrium: an alternative to phyletic gradualism." Pp. 82–115 in T. Schopf, ed., *Models in Paleobiology*. San Francisco: Freeman Cooper.
Fagan, B. 2001. *The Little Ice Age: How Climate Made History, 1300–1850*. Boston, MA: Basic Books.
Fagan, B., and N. Durrani. 2016. *A Brief History of Archaeology*. London: Routledge.
Feder, K. 2010. *Encyclopedia of Dubious Archaeology*. Santa Barbara, CA: Greenwood.
Fernandez, F. 2015. "Human dispersal and late quaternary megafaunal extinctions: the role of the Americas in the global puzzle." P. 56 in *A Genetic and Biological Perspective of the First Settlements of the Americas*. Mexico City: UNESCO.
Firestone, R., J. West, J. Kennett, et al. 2007. Evidence for an extraterrestrial impact 12,9000 years ago that contributed to the megafaunal extinctions and the Younger Dryas cooling. *Proceedings of the National Academy of Sciences* 104: 16016–16021.
Flannery, T. 2018. *Europe: A Natural History*. New York: Atlantic Monthly Press.
Fleagle, J. 1999. *Primate Adaptation and Evolution*. San Diego: Academic Press.
Fleming, J. 1998. *Historical Perspectives on Climate Change*. New York: Oxford University Press.
Fleming, J. 2006. James Croll in context: The encounter between climate dynamics and geology in the second half of the nineteenth century. *History of Meteorology* 3: 43–54.
Flores, D. 2016. *American Serengeti*. Lawrence: University of Kansas Press.
Fortey, R. 2000. *Tribobite! Eyewitness to Evolution*. New York: Alfred A. Knopf.

Friedman, R. 2003. *The Bible with Sources Revealed.* New York: HarperCollins.

Gere, C. 2011. *The Tomb of Agamemnon.* Cambridge: Harvard University Press.

Geikie, J. 1874. *The Great Ice Age, and Its Relation to the Antiquity of Man.* New York: D. Appleton and Company.

Geikie, J. 1894. *The Great Ice Age, and Its Relation to the Antiquity of Man, 3rd edition.* London: Edward Stanford.

Gertner, J. 2019. *The Ice at the End of the World.* New York: Random House.

Gibbons, A. 2019. Ancient jaw gives elusive Denisovans a face. *Science* 364: 418–419.

Gilligan, D. 2011. *Rise of the Ranges of Light.* Berkeley, CA: Heyday.

Godinot, M. 2000. Rafting on a wide and wild ocean. *Science* 368: 194–95.

Golding, W. 1955. *The Inheritors.* London: Faber & Faber.

Goodall, J. 1986. *The Chimpanzees of Gombe.* Cambridge, MA: Harvard University Press.

Gould, S. 2002. *The Structure of Evolutionary Theory.* Cambridge, MA: Harvard University Press.

Grabau, A., and H. Shimer. 1909. *North American Index Fossils.* New York: A.G. Seiler.

Grabau, A. 1921. *A Textbook of Geology. Part II. Historical Geology.* Boston: D.C. Heath & Co.

Grabau, A. 1924. *Principles of Stratigraphy.* New York: A.G. Seiler. Republished with introduction by M. Kay, 1960. New York: Dover.

Grabau, A., and H. Schimer. 1909. *North American Index Fossil: Invertebrates.* New York: A.G. Seiler.

Graham, R., S. Belmecheri et al. 2016. Timing and causes of mid-Holocene mammoth extinction at St. Paul Island, Alaska. PNAS 113: 9310–9314. doi: 10.1073/pnas.1604903113

Grayson, D. 2016. *Giant Sloths and Sabertooth Cats.* Salt Lake City: University of Utah Press.

Greenberg, D. 1999. *The Politics of Pure Science.* Chicago: University of Chicago Press.

Gribbin, J., and M. Gribbin. 2015. *Ice Age: The Theory That Came in from the Cold.* Golden CO: ReAnimus Press.

Gunn, R. et al. 2011. What bird is that? *Australian Archaeology* 73: 1–12.

Guthrie, R. 2001. Origin and causes of the mammoth steppe: A story of cloud cover, woolly mammal tooth pits, buckles, and inside-out Beringia. *Quaternary Science Reviews* 20: 549–74.

Haeckel, E. 1879. *The Evolution of Man.*

Haigh, J. 2007. The sun and the earth's climate. Living Review in Solar Physics 4:2. doi.ief/10,12942/lrso-2007-2

Handwerk, B. 2020. A mysterious 25,000-year-old structure built of the bones of 60 mammoths. *Smithsonian Magazine* (Mar. 16), smithsonianmag.com.

Hartmann, J. 2019. "Plate tectonics, carbon, and climate." *Science* 364: 126–127. doi: 10.1126/science.aax1657

Hays, J. J., Imbrie and N. Shackleton. 1976. Variations in the Earth's orbit: Pacemaker of the ice ages. *Science* 194: 1121–1132.

Haeckel, E. 1904. *Art Forms in Nature*. Munich: Prestel.

Hodder, I. 2011. *The Leopard's Tale: Revealing the Mysteries of Catalhoyuk*. London: Thames & Hudson.

Hodson, A., et al. 2008. "Glacial ecosystems." Ecological Monographs 78. doi .org/10.1890/07-0187.1

Hodge, M. 1865. *Hans Brinker, or The Silver Skates*. New York: F.O.C. Darley and Thomas Nast.

Hoffmann, D., et al. 2018. U-Th dating of carbonate crusts reveals Neandertal origin of Iberian cave art. *Science* 359: 912–15.

Holdren, C. 1998. No last word on language origins. *Science* 282: 1455–59.

Hull, D., P. Tessner, and A. Diamond. 1978. Planck's Principle. *Science* 202: 717–23.

Hull, P., et al. 2020. On impact and volcanism across the Cretaceous-Paleogene boundary. *Science* 367: 266–71.

Hutton, J. 1788. *Theory of the Earth*. Edinburgh: Transactions of the Royal Society of Edinburg.

Hyatt, A. 1897a. Cycle in the Life of the Individual (Ontogeny) and in the Evolution of Its Own Group (Phylogeny). *Proceedings of the American Academy of Arts and Sciences*. 32: 209–24.

Imbrie, J., and K. Imbrie. 1979. *Ice Ages*. Cambridge, MA: Harvard University Press.

IPCC. 2001. *Climate Change 2001: Working Group I: The Scientific Basis*. World Meteorological Organization, Geneva, Switzerland.

IPCC. 2021. *Climate Change 2021: The Physical Science Basis*. Contribution of Working Group I to the Sixth Assessment Report of the Intergovernmental Panel on Climate Change [Masson-Delmotte, V., P. Zhai, A. Pirani, S. L. Connors, C. Péan, S. Berger, N. Caud, Y. Chen, L. Goldfarb, M. I. Gomis, M. Huang, K. Leitzell, E. Lonnoy, J. B. R. Matthews, T. K. Maycock, T. Waterfield, O. Yelekçi, R. Yu and B. Zhou (eds.)]. Cambridge University Press. In Press.

Irving-Pease, E., R. Hannah, A. Jamieson, et al. 2018. Paleogenomics of animal domestication. Pp. 225–72 in C. Lindqvist and O. Rajora (eds.), *Paleogenomics. Population Genomics*. Springer. doi: 10.10.1007/13836_2018_2018_55.

Jackson, R. 2019. Eunice Foote, John Tyndall and a question of priority. Notes Rec. 74: 105–118. Doi.org/10.1098/rsnr.2018.0066

Jaffe, M. 2001. *The Gilded Dinosaur: The Fossil War between E.D. Cope and O.C. Marsh and the Rise of American Science*. New York: Three Rivers Press.

Johanson, D., and K. Wong. 2009. *Lucy's Legacy: The Quest for Human Origins*. New York: Crown Publishing.

Jordan, P. 1999. *Neanderthal*. Gloucestershire UK: Sutton Publishing.

Jouzel, J. 2013. "A brief history of ice core science over the last 50 years." *Climate of the Past* 9: 2525–47. doi: 10.5194/cp-9-2525—2013

Jouzel, J., V. Masson-Delmotte, O. Cattani, et al. 2007. Orbital and millennial Antarctic climate variability over the past 800,000 years. *Science* 317: 793–796. doi: 10.1126/science.1141038

Kennett, J., D. Kennett, B. Culleton, et al. 2015. Bayesian chronological analyses consistent with synchronous age of 12,835-12,735 B.P. for Younger Dryas boundary on four continents. *PNAS* 112: E4344–E4353. doi: 10.1073/pnas,1507146112

Kjaer, K, et al. 2018. A large impact crater beneath Hiawatha Glacier in northwest Greenland. *Science Advances* 4: eaar8173.s. doi: 10.1126/sciadv.aar8173

Koestler, A. 1964. *The Sleepwalkers*. New York: Oxford University Press.

Kolbert, E. 2014. *The Sixth Extinction*. New York: Henry Holt.

Kozowyk, P., M. Soressi, D. Pomstra et al. 2017. Experimental methods for the Palaeolithic dry distillation of birch bark: Implications for the origin and development of Neandertal adhesive technology. *Scientific Reports* 7: 8033. doi: 10.1038/s41598–017-08106-7

Köppen, W., and A. Wegener. 1924. *Climates of the Geological Past*. Berlin: Borntraeger.

Kramer, J. 2002. *The Art of Flowers*. New York: Watson-Guptill.

Krüger T. 2013. *Discovering the Ice Ages: International Reception and Consequences for a Historical Understanding of Climate*. Leiden: Brill.

Lanham, U. 2011. *The Bone Hunters: The Heroic Age of Paleontology in the American West*. Mineola, NY: Dover Publications.

Leakey, L., P. Tobias, and J. Napier. 1964. *Nature* 202: 7–9.

Leakey, R., and R. Lewin. 1996. *The Sixth Extinction*. London: Weidenfeld & Nicolson.

Lehmann, J. 1756. *Versuch einer Geschichte des Flötz-Gebűrgen*. Berlin.

Lisiecki, L., and M. Rayno. 2005. A Pliocene-Pleistocene stack of 57 globally distributed benthic O-18 records. *Paleoceanography* 20 PA1003. doi: 10.1029/2004PA001071

Liu, L., and X. Chen. 2012. *The Archaeology of China: From the Late Paleolithic to the Early Bronze Age*. Cambridge, UK: Cambridge University Press.

Lyell, C. 1834. *Principles of Geology*, 3rd edition. London: John Murray.

Lyell, C. 1835. *Principles of Geology*, Volume 1 (4th edition). London: John Murray.

Lycett, S., and J. Gowlett. 2008. On questions surrounding the Acheulean "tradition." *World Archaeology* 40: 295–315. doi: 10.1080/00438240802260970

Lyson, T., I. Miller, et al. 2019. Exceptional continental record of biotic recovery after the Cretaceous–Paleogene mass extinction. *Science* 24: eaay2268. doi: 10.1126/science.aay2268

Macdonald, F., N. Swanson-Hysell, Y. Park et al. 2019. Arc-continent collisions in the tropics set Earth's climate state. *Science* 364: 181–84.

MacLeod, N. 2013. *The Great Extinctions*. London: Natural History Museum.

MacPhee, R. 2019. *End of the Megafauna*. New York: W.W. Norton.

Mann, M. 2012. *The Hockey Stick and the Climate Wars*. New York: Columbia University Press.

Mann, C. 2006. *1491: New Revelations of the Americas before Columbus*. New York: Random House.

Mann, C. 2012. *1493: Uncovering the New World Columbus Created*. New York: Random House.

Mapes Dodge, Mary. 1865. Hans Brinker; or, *The Silver Skates: A Story of Life in Holland*. New York: James O'Kane.

Martin, P. 2005. *Twilight of the Mammoths: Ice Age Extinctions and the Rewilding of America*. Berkeley, CA: University of California Press.

Martin, R. D. 1990. *Primate Origins and Evolution*. Princeton, NJ: Princeton University Press.

Matsuzawa, T. 2001. *Primate Origins of Human Cognition and Behavior*. Hong Kong: Springer.

Mazur, A. 2004. *A Romance in Natural History: The Lives and Works of Amadeus Grabau and Mary Antin*. Syracuse, NY: Garret.

Mazur, A. 2005. *Biosociology of Dominance and Deference*. New York: Rowman & Littlefield.

Mazur, A. 2007. *Global Social Problems*. Lanham, MD: Rowman & Littlefield.

Mazur, A. 2008. *Implausible Beliefs in the Bible, Astrology and UFOs*. New Brunswick, NJ: Transaction Publishers.

Mazur, A. 2009. *The Female Nude in Western Art*. Syracuse, NY: Garret.

Mazur, R. 2015. *Speaking of Bears: The Bear Crisis and a Tale of Rewilding from Yosemite, Sequoia, and Other National Parks*. Helena, MT: Falcon Guides.

Mazur, A., and L. Robertson. 1972. *Biology and Social Behavior*. New York: Free Press.

McClellan, J., and H. Dorn. 2006. *Science and Technology in World History: An Introduction*. Baltimore: Johns Hopkins University Press.

McPherron, S., et al. 2010. Evidence for stone-too-assisted consumption of animal tissues before 3.39 million years ago at Dikika, Ethiopia. *Nature* 466: 857. doi: 10.1038/Nature09248

Mellars, P. 2009. Origins of the female image. *Nature* **459**: 176–177. doi: org/10.1038/459176a

Meltzer, D. 2015. *The Great Paleolithic War: How Science Forged an Understanding of America's Ice Age Past*. Chicago: University of Chicago Press.

Milankovitch, M. 1920.

Milankovitch, M. 1938.

Mithen, Steven. 2003. *After the Ice: A Global Human History 20,000–5000 BC*. Cambridge, MA: Harvard University Press.

Montes, C., A. Cardona, C. Jaramillo, A. Pardo, J. Silva, and V. Valencia. 2015. Middle Miocene closure of the Central American seaway. *Science* 348: 226-229. doi: 10.1126/science.aaa2815

Mooallem, J. 2017. Neanderthals were people, too. *New York Times Magazine* Jan. 11.

Morris, S. 2003. *Life's Solution*. New York: Cambridge University Press.

Murowchick, R. (ed.). 1994. *Cradles of Civilization-China: Ancient Culture, Modern Land*. Norman OK: University of Oklahoma Press.

Myrow, P., M. Lamb, and R. Ewing. 2018. Rapid sea level rise in the aftermath of a Neoproterozoic snowball Earth. *Science* 360: 649–651.

National Research Council. 1982. *Climate in Earth History: Studies in Geophysics*. Washington, DC: The National Academies Press. doi: ord/10.17226/11798

Needham, J. 1986. *Science and Civilization in China: Volume 4, Physics and Physical Technology*. Taipei: Caves Books, Ltd.

Newman, R. 1995. American intransigence: The rejection of continental drift in the great debates of the 1920s. *Earth Sciences History* 14: 62–83.

O'Brien, M., B. Buchanan, and M. Eren (eds.). 2018. *Convergent Evolution in Stone-Tool Technology*. Cambridge, MA: MIT Press.

O'Dea, A., H. Lessios, et al. 2016. "Formation of the Isthmus of Panama." *Science Advances* 2: e1600883.

O'Neill, D. 2004. *The Last Giant of Beringia*. New York: Basic Books.

Oreskes, N. 1999. *The Rejection of Continental Drift*. New York: Oxford University Press.

Osiurak, F., and E. Reynaud. 2020. The elephants in the room: What matters cognitively in cumulative technological culture. *Behavioral and Brain Sciences*.

Pääbo, S. 2015. *Neanderthal Man: In Search of Lost Genomes*. New York: Basic Books.

Parker, G. 2013. *Global Crisis: War, Climate Change and Catastrophe in the Seventeenth Century*. New Haven CT: Yale University Press.

Pascoe, B. 2018. *Dark Emu: Aboriginal Australia and the Birth of Agriculture*. London: Scribe Publications.

Perri, A., T. Feuerborn, L. Frantz, G. Larson, R. Malhi, D. Meltzer, and K. Witt. 2021. Dog domestication and the dual dispersal of people and dogs into the Americas. *PNAS* 118 (6): e2010083118. doi: 10.1073/pnas.2010083118

Peters, S., and R. Gaines. 2012. Formation of the "Great Unconformity" as a trigger for the Cambrian explosion. *Nature* 484: 363–366.

Petit, J. R., I. Basile, A. Leruyuet et al. 1997. Four climate cycles in Vostok ice core. *Nature* 387: 359–360.

Petit, J. R., J. Jousel, D. Raynard et al. 1999. Climate and atmospheric history of the past 420,000 years from the Vostok ice core, Antarctica. *Nature* **399**, 429–436 (1999).

Pfeiffer, J. 1982. *The Creative Explosion*. Ithaca, NY: Cornell University Press.

Phillips, J. 1841. *Figures, and Descriptions of the Palaeozoic Fossils of Cornwall, Devon, and West Somerset*. London: Longman, Brown, Green, & Longmans.

Phillips, J. 1860. *Life on the Earth: Its Origin and Successions*. London: Macmillan.

Pierce, P. 2006. – *Jurassic Mary: Mary Anning and the Primeval Monsters*. The History Press. Gloucestershire UK.

Pinker, S. 1994. *The Language Instinct*. New York: William Morrow and Company.

Pollan, M. 2006. *The Omnivore's Dilemma*. London: Penguin Books.

Pomeroy, E., P. Bennett, C. Hunt et al. 2020. New Neanderthal remains associated with the "flower burial" at Shanidar Cave. *Antiquity* 94: 11–16. doi: 10.15185/aqy.2019.207

Prothero, D. 2017. *The Princeton Field Guide to Prehistoric Mammals*. Princeton: Princeton University Press.

Pryor, A., D. Beresford-Jones, A. Dudin et al. 2020. The chronology and function of a new circular mammoth-bone structure at Kostenki 11. *Antiquity*. doi: org/ 10.15184/agy.2020.7

Rasmussen, S., K. Andersen et al. 2006. A new Greenland ice core chronology for the last glacial termination. *Journal of Geophysical Research* 111 (D6): D06102. doi: 10.1029/2005JD006079.

Raup, D. 1995. The Role of Extinction in Evolution. P. 6 in W. Fitch and F. Ayala (Eds), *Tempo and Mode in Evolution*. Washington, DC: National Academies Press.

Raymo, C., and M. Raymo. 2007. *Written in Stone: A Geological History of the Northeastern United States*. Delmar, NY: Black Dome Press.

Reich, D., et al. 2010. Genetic history of an archaic hominin group from Denisova Cave in Siberia. *Nature* 468: 1053–1060. doi: 10.1038/nature09710

Reich, D. 2018. *Who We Are and How We Got Here*. New York: Pantheon Books.

Revelle, R., and H. Suess. 1957. Carbon dioxide exchange between atmosphere and ocean and the question of an increase of atmospheric CO_2 during the past decades. *Tellus* 9: 18–27.

Rich, N. 2019. *Losing Earth: A Recent History*. New York: Macmillan.

Richerson, P., R. Boyd, and R. Bettinger. 2002. Was agriculture impossible during the Pleistocene but mandatory during the Holocene? A climate change hypothesis. *American Antiquity* 66: 387–411.

Richerson, P., S. Gavrilets, and F. de Waal. 2021. Modern theories of human evolution foreshadowed by Darwin's Descent of Man. *Science* 372: eaba3776. doi: 10.1126/science.aba3776

Rizal, Y., J. Westaway, Y. Zaim, et al. 2019. Last appearance of *Homo erectus* at Ngandong, Java, 117,000-108,000 years ago. *Nature.* doi: 10.1038/s41586–019-1863-2

Rudwick, M. 2008. *Bursting the Limits of Time: The Reconstruction of Geohistory in the Age of Revolution.* Chicago: University of Chicago Press.

Rudwick, M. 2014. *Earth's Deep History: How It Was Discovered and Why It Matters.* Chicago: University of Chicago Press.

Sandom, C., S. Faurby, B. Sandel et al. 2014. Global late Quaternary megafauna extinctions linked to humans, not climate change. *Proc. Royal Society B* 281: 20133254.

Saltré, F., et al. 2016. Climate change not to blame for late Quaternary megafauna extinctions in Australia. *Nature Communications* 7: 10511. doi: 10.1038/ncomms10511/www.nature.com/naturecommunications

Sahlins, M. 1968. Notes on the original affluent society. Pp. 85–89 in R. Lee and I. DeVore (eds.), *Man the Hunter.* Chicago: Aldine.

Scarre, C. 2018. *The Human Past: World Prehistory and the Development of Human Societies.* London: Thames & Hudson Ltd.

Scott, J. 2017. *Against the Grain.* New Haven: Yale University Press

Scott, W. 1913. *A History of Land Mammals in the Western Hemisphere.* New York: MacMillan Publishing Company.

Secord, J. 2000. *Victorian Sensation.* Chicago: University of Chicago Press.

Shea, J. 2017. *Stone Tools in Human Evolution.* New York: Cambridge University Press.

Shelley, 1818. *Frankenstein; or, The Modern Prometheus.* London: Lackington, Hughes, Harding, Mavor & Jones.

Shipman, P. 2001. *The Man Who Found the Missing Link.* New York: Simon & Schuster.

Shipman, P. 2015. How do you kill 86 mammoths? Taphonomic investigations of mammoth megasites. *Quaternary International* 359–360L: 38–46. doi: 10.1016/j.quaint.2014.04.048

Silverberg, R. 1997. *Great Adventures in Archaeology.* Lincoln: University of Nebraska Press.

Simpson, G. 1951. *Horses.* New York: Oxford University Press.

Singh, U. 2009. *A History of Ancient and Early Medieval India.* Delhi: Pearson.

Slao, V. et al. 2018. The genome of the offspring of a Neanderthal mother and a Denisovan father. *Nature.* doi.org/10.1038.

Smith, M. 2019. *Cities.* London: Simon & Schuster.

Soffer, O. 1985. *The Upper Paleolithic of the Central Russian Plain.* London: Academic.

Sutikna, T. et al. 2016. Revised stratigraphy and chronology for *Homo floresiensis* at Liang Bua in Indonesia. *Nature* 532: 366–369. Bibcode:2016Natur.532.366S.

Swisher, III, C., G. Curtis, and R. Lewin, 2000. *Java Man*. Chicago, IL: University of Chicago Press.

Telford, T., and H. Shaban. 2019. "You can buy a baby T. rex skeleton on eBay for $3 million. Scientists would rather you didn't." *Washington Post* (Apr. 19).

Theunissen, B. 1998. *Eugene Dubois and the Ape-Man from Java*. Boston, MA: Kluwer Academic Publishers.

Trigger, B. 2001. *Understanding Early Civilizations*. Cambridge, UK: Cambridge University Press.

Trigger, B. 2003. *Understanding Early Civilizations*. Cambridge, UK: Cambridge University Press.

Tuck, J., and R. Grenier. 1985. "Discovery in Labrador: A 16th-century Basque whaling port." *National Geographic* 168 (July): 40–68.

Tyndall, J. 1864. On radiation through the earth's atmosphere. *Journal of the Franklin Institute* 77: 413–418.

Vartanyan, D. 1995. Radiocarbon dating evidence for mammoths on Wrangel Island, Arctic Ocean, until 2000 BC. *Radiocarbon* 37: 1–6. doi: 10.1017/S0033822200014703

Veizer, J., et al. 1999. $^{87}Sr/^{86}Sr$, $d^{13}C$ and $d^{18}O$ evolution of Phanerozoic seawater. *Chemical Geology* 161: 59–88. doi: 10.1016/50009-2541(99)00081-9

Voosen, P. 2018. Ice age impact. *Science* 362: 738–742.

Voosen, P. 2019. Project traces 500 million years of roller-coaster climate. *Science* 364: 716–717.

Walker, J., P Hays and J. Kasting. 1981. A negative feedback mechanism for the long-term stabilization of Earth's surface temperature. *Journal of Geophysical Research: Oceans* 86: 9776-9782. doi: 10.1029/JC086iC10p09776

Waters, M. 2019. Late Pleistocene exploration and settlement of the Americas by modern humans. *Science* 365. doi: 10,1126/science.aat5447

Weart, S. 2008. *The Discovery of Global Warming: Revised and Expanded Edition*. Cambridge, MA: Harvard University Press.

Wilkinson, T. 2010. *The Rise and Fall of Ancient Egypt*. London: Bloomsbury Publishing.

Wilkinson, T. 2020. *A World Beneath the Sands*. London: W. W. Norton.

Will, M. 2020. Neanderthal surf and turf. *Science* 367: 1422–1423.

Willeit, M., A. Ganopolski, R. Calov et al. 2019. Mid-Pleistocene transition in glacial cycles explained by declining CO_2 and regolith removal. *Science Advances* 5: eaav7337.

Williamson, C., K. Cameron, J. Cook et al. 2009. Glacial algae: A dark past and a darker future. Frontiers in Microbiology (Apr. 4). doi: 10.3389/fmicb.2019.00524

Wilson, G. 2008. A wild gift from the West: Two grizzly cubs. *Monticello Newsletter* 19.

Wilson, J. 1966. "Did the Atlantic close and then re-open?" *Nature* 211: 676-681.

Winchester, S. 2001. *The Map That Changed the World*. New York: Penguin Group.

Witmer, L. 1995. The extant phylogenetic bracket and the importance of reconstructing soft tissues in fossils. In *Functional Morphology in Vertebrate Paleontology*, ed. J. Thomasson, 19–33. New York: Cambridge University Press.

Wojtal, P, J. Wilczynski, and G. Haynes. 2015. World of Gravettian Hunters. *Quaternary International* 350–360: 1–2. doi: 10.1016/j.quaint.2014.12.043

Wood, B. 2014. Fifty years after *Homo habilis*. *Nature* 508: 31–33.

Woolley, L. 1955. *Spadework; Adventures in Archaeology*. Cambridge, UK: Lutterworth Press.

Woodward, J. 2014. *The Ice Age: A Very Short Introduction*. New York: Oxford University Press.

Wragg Sykes, R. 2020. *Kindred*. London: Bloomsbury Sigma.

Wright, R. 2009. *The Ancient Indus: Urbanism, Economy, and Society*. Cambridge, UK: Cambridge University Press.

Wunsch, C. 2002. What Is the thermohaline circulation? *Science* 298: 1179–1181. doi: 10.1126/science.1079329

Yang, M., A. Fam, B. Sun, et al. 2020. Ancient DNA indicates human population shifts and admixture in northern and southern China. *Science* 369: 282–288. doi: 10.1126/science.aba0909

Zahid, J., E. Robinson, and R. Kelly. 2016. Agriculture, population growth, and statistical analysis of the radiocarbon record. *PNAS* 113: 931–936. doi: 10/1073/pnas.1517650112

Zong, Y., Z. Chen, J. Innes et al. 2007. Fire and flood management of coastal swamp enabled first rice paddy cultivation in each China. *Nature* 449: 459–463. doi: 10.1038/nature06135

INDEX